IEE ELECTROMAGNETIC WAVES SERIES 2

SERIES EDITORS: J.R. WAIT, G. MILLINGTON AND E.D.R. SHEARMAN

Electromagnetic waves and curved structures

Electromagnetic waves and curved structures

LEONARD LEWIN,
D.Sc., Sen. Mem. I.E.E.E., F.B.I.S, C.Eng., M.I.E.E.
Professor of Electrical Engineering University of Colorado

DAVID C. CHANG,
Ph.D., Mem. I.E.E.E.
Professor of Electrical Engineering University of Colorado

EDWARD F. KUESTER,
M.S., Ph.D., Mem. I.E.E.E.
Assistant Professor of Electrical Engineering University of Colorado

PETER PEREGRINUS LTD.
On behalf of the Institution of Electrical Engineers

Published by Peter Peregrinus Ltd.
Southgate House, Stevenage, Herts. SG1 1HQ, England

ISBN: 0 901223 96 4

Typeset by Alden Press., London and Northampton
Printed in England by A. Wheaton & Co. Ltd., Exeter

Contents

Introduction

Although the propagation of radio waves round the Earth is one of the most important examples of the interaction of electromagnetic waves with a curved structure, it is a rather special subject with a large and growing literature (see, for example, the book by Wait[1] which also contains many references), and will not be further referred to in this monograph. We shall be concerned instead with propagation in, and radiation from, closed and open guiding structures of both metallic and dielectric construction.

Perhaps the earliest such work on curved metallic structures was that of Pocklington,[2] who studied the current propagation on curved wires of small cross-section and applied his results to the closed ring and helix configurations: these early formulas for helix propagation pre-dated the later and similar work on travelling-wave helix structures by some fifty years.

The examination in the thirties of the wave-guiding properties of metallic pipes led, in 1939, to what was probably the first substantial investigation of propagation in curved waveguides. Buchholz,[3] using a Bessel-function formulation, studied the propagation round curved rectangular waveguides viewed as sectors of a cylindrical space. The big problem with this Bessel formulation, which in principle is exact, is that subsequent equations have to be solved for the Bessel-function order, and no exact analytic solutions are known. Buchholz utilized a number of mathematically interesting but lengthy and highly involved formulas to obtain some approximate solutions. More recently, using Debye asymptotic expansions, the same solutions have been obtained rather more readily by later authors.

Marshak[4] in 1943 calculated both the curved guide propagation factors and a formula for the reflection at the junction of a straight and curved rectangular waveguide, obtaining the leading terms in an expansion for large radius of curvature. This unpublished wartime report is not too well-known, and in the late forties and early fifties several authors came out with similar and independent calculations, in which the key feature was the expression of Maxwell's equations in local co-ordinate form and the development of expansions in inverse powers of the curvature radius. One of the better known sets of results of this perturbation process is due to Jouguet,[5–7] who produced in 1946 and 1947 a series of papers utilizing this method. They cover propagation in curved rectangular waveguide; junction effects obtained via local mode expansions; and propagation in curved circular pipes, particularly the important case of TE_{01} and TM_{11} mode coupling. About the same time Riess[8] examined the propagation of general-order modes in curved rectangular waveguide, again using asymptotic expansion methods. He sought to describe the modes by the equivalent of the TE and TM modes; but unfortunately in the curved section this form of selection is not appropriate, and he came to the erroneous conclusion that exact solutions involving cylindrical functions would not be possible. In a subsequent note on this paper Schelkunoff[9] showed that if a different mode structure were selected, viz. the longitudinal electric or magnetic modes (denoted by LE and LM), then an exact Bessel-type solution is indeed possible; though, as already stated, the extraction of information from this mathematical formulation is not too straightforward. All subsequent writers appear to have utilized this LE and LM mode formulation.

About the same time as these asymptotic methods were being explored a completely different transmission line approach, initially by researchers at the Bell Laboratories, was yielding interesting results. In 1948, Rice[10], in a classic paper, used a matrix formulation of the generalized telegraphist's equations to obtain an extremely general set of equations for a guide of arbitrary cross-section with curved axis. The matrix elements involved mode-coupling coefficients which were obtained as rather general integrals of the mode basis functions and the space variables. Gentle bends give rise to near-diagonal matrices and some useful formulas for these are given. Asymptotic forms of the coupling coefficients for gentle bends are also determined and applied to propagation in rectangular guide bent into an arc of a circle between two sections of straight guide. The results include the junction reflection and the propagation coefficients as the first terms in a series of inverse powers in the radius of curvature of the bend. For very sharp bends the matrix equations are tackled directly and results calculated numerically by truncation of the matrices at third order. By comparison with truncation at first and second order it seems that third order truncation should yield accurate numerical results. A comparison with the results for 'gentle' bends shows that, for a 90° bend, the asymptotic results are quite good even for radii of curvature of the guide center line as small as the guide width; giving a rather remarkable range of usefulness.

A somewhat similar approach by Schelkunoff[11] in 1955 converts Maxwell's equations into the generalized telegraphist's equations. The form of the coupling coefficients is obtained, though the equations themselves are not solved in the paper. Applications are to a parallel plate horn of arbitrary shape or a bent waveguide. The basis functions use rectangular coordinates for the most part, but one example in curvilinear co-ordinates is given.

Both these papers are very lengthy, despite the omission of much detail in many places. Because the originals are generally available and provide a very adequate treatment of the method it was decided to omit further reference to them in this monograph, which is mainly concerned with the perturbation approach. However, the transmission line methods seem better adapted to numerical work involving *very* sharp curvatures, and may be superior in dealing with some aspects of variable curvature.

A somewhat different approach by Albersheim[12] in 1949 dealt with the problem of curvature in a circular guide carrying the TE_{01} mode by comparison with formulas for coupled circuits. Useful results on bending were obtained in this way, including the effect of the imperfect conductivity of the guide wall in breaking the degeneracy between the TE_{01} and TM_{11} modes.

Propagation along a helix has already been mentioned. In the forties and fifties many studies of this structure took place, most in relation to travelling-wave amplifier circuits. The 'sheath' helix was a popular model and much work was done on it, but it really does not fit into the present context. In contrast, Sollfrey[13] in 1951 set up the general form of the Maxwell equations in the (non-orthogonal) helix system and proceeded to an approximate solution of the case of small wire radius. His general equations, for large helix radius, would apply to radiation for a helical antenna but additional work would be needed to get useful analytical solutions from the basic equations.

Lewin[14] in 1955 applied the perturbation technique to curved and twisted waveguide of rectangular cross-section and developed a non-orthogonal twisted co-ordinate system with rectangular boundary surfaces for the treatment of the latter. Waldron[15] in 1957 approximately treated the case of a rectangular guide whose axis was bent into a tight helical path by deforming a guide in a circular path.

A quite different approach to propagation in gently curved guides was given in 1956 in a paper by Sveshnikov.[16] Maxwell's equations were presented in a form similar to the straight guide form plus correction terms depending on the curvature. These additional terms were then interpreted as sources and the problem solved iteratively through a Green's function formulation. This method is capable of dealing with either variable or constant curvature and gives results equivalent to the previous methods.

The generalized telegraphist's equations were used by Morgan[17] in 1957 and Quine[18] in 1965. Morgan treated the case of a variable dielectric in curved circular guide, and sought configurations of dielectric which would permit bending with minimum mode conversion. Quine, using only the forward-coupled waves with the coupling coefficients from Schelkunoff's paper examined mode-coupling in overmoded rectangular guide intended for very high power application. Numerical integration was used to obtain the coupled-mode content at the output of a circular bend. Sinusoidal bends were also examined by using a curvature-dependent variable coupling in the equations.

In 1966 Cochran and Pecina[19] investigated mode propagation in curved guide by tackling numerically solutions for the order of the Bessel-function formulation. They also obtained analytic solutions for the two extremes of gradual bends and very high frequencies (far from cutoff), and investigated the ranges of validity of each approximation. This paper contains also an excellent bibliography.

The asymptotic perturbation approach has features in common with ray-optical techniques, whose application to waveguide problems can be found in a sizeable body of literature on the topic. We mention here only the papers by Burstein and Solov'ev[20] in 1955, and more recently, those of Maurer and Felsen[21] in 1970 and Ahluwalia, Keller and Matkowsky[22] in 1974. The first of these took the Maxwell equations in curvilinear co-ordinates and adapted them to the solution of guided propagation between gently curved parallel double-curvature surfaces. The phase and ray-path solutions were found using perturbation techniques. The review paper of Maurer and Felsen, as well as the more recent paper, discusses waveguide propagation from the viewpoint of ray congruences, generalizing a well-known interpretation of modes in straight guides. The latter paper also presents a 'boundary layer' analysis which is closely related to the perturbation solutions of Jouguet[5–7] and Lewin,[14] as well as a uniform asymptotic solution corresponding to the 'high frequency' case studied by Cochran and Pecina.[19] The ray method has proved particularly useful for heavily overmoded guides.

Two papers, by Bahar[23] in 1969, and Bahar and Govindarajan[24] in

1973 showed a fresh approach to the subject of curved waveguides. Essentially the new feature comes from recognizing that an annular basis (Bessel-function approach) is a natural mode for the uniformly curved guide, whereas the rectangular basis (trigonometric functions as used in the coupled transmission line approach) is not. Hence the latter, in a region of uniform curvature, gives rise to coupled modes which the former does not. In the case of variable bends the important coupled modes are much fewer, and are more closely bunched around the main mode, in the case of the annular basis. The coupling coefficients, in both formulations, are obtained by field matching at inclined elementary waveguide sections. An example of a sinusoidal bend is treated numerically and exemplifies the above characteristics.

Finally, in the context of curved metallic waveguides, we would mention the 1969 paper of Bates[25] who uses a junction-aperture integral equation approach to solve the problem of a junction between straight and curved rectangular guide-sections. A set of annular basis functions is used in the curved region and the integral equation is solved numerically by moment methods and truncation. Many typical results are given for mode coupling, in the form of graphs and tables. In particular a sharp increase of the junction reflection coefficient in the immediate neighborhood of cut-off is shown. The reason for this phenomenon is traced to the fact that the straight guide can be propagating whilst the curved guide may be just beyond cutoff; or vice versa, depending on the polarization in relation to the plane of the bend. The same feature was remarked on by Jouguet[6] in one of the above-mentioned earlier papers.

In all the preceeding research, interesting and useful analytic techniques are developed as part of the process of solving the problems in hand. During the period a number of papers of a general character relating to propagation in curved waveguides of constant cross-section also appeared. Tang[26] in 1970 showed that, in general, non-orthogonal co-ordinate systems will ensue when Maxwell's equations are adapted to a curved axis, due to the torsion of the axial path. By introducing a derived coordinate system that twists around the original to give zero net torsion, an orthogonal system, with consequent simplifications, can result. Implicit in this approach is the corresponding twisting of the boundary surfaces to conform with the new co-ordinates. Of course, if this twisting is undesirable in any application, the method would not be suitable. The non-orthogonal twisted co-ordinate system[14] introduced to investigate propagation in twisted rectangular guide could be seen as an exemplification of this – Tang's approach would merely have untwisted the guide! In a sense, the circular guide is an exception since the twisting leaves its shape unaltered. However, the co-ordinate systems, and with it the basis for representing the waveguide field pattern, twists inside the guide, though this may be of no practical consequence from an application point of view.

Orthogonal or not, Maxwell's equations in curvilinear co-ordinates are separable only in special instances. A general approach to handling them for waveguide applications was considered by Lewin[27] in 1972, using bi-complex variables to decouple the field components. The final equations are then in a suitable form for perturbation analysis or solution by iteration.

So far the discussion has referred to closed metallic structures. However, open dielectric guides assumed a prominence in the late sixties particularly in relation to their potential for optical transmission. Both the dielectric slab and the cylindrical configuration are of importance, and analyses of the straight configurations had long been classical. That curvature of such guides introduces radiation was apparently first demonstrated (as a corollary to the proof that dielectric resonators must radiate) by Richtmyer[28] in 1939. Richtmyer also presented formulas for the radiative attenuation of both slabs and cylindrical fibers which include the salient features of this phenomenon. Later, analyses dealing with similar problems arising in radio wave propagation around the Earth appeared, but not until Marcatili's[29] 1969 analysis of curved slabs and rectangular channels was specific attention paid to optical waveguide structures.

While Marcatili solved the slab problem by directly perturbing the exact eigenvalue equation obtained by field matching at the boundaries, Marcuse[30] in 1971 performed a far-field Poynting-vector integration to calculate the radiated power. Maurer and Felsen[21] also tackled the slab problem, as did Chang and Barnes[31] in 1973. The latter, using what amounted to a perturbation solution of the exact eigenvalue equation, showed that a shift in the modal phase velocity introduced by the curvature could have a substantial effect on the radiation, and showed how partial reduction of this loss could be achieved by intentionally changing the phase velocity. In 1975, slabs with more general refractive index distributions were considered by Heiblum and Harris[32] and by Kawakami, Miyagi and Nishida,[33] using a WKB method as was done by Chang and Barnes. These two papers also consider the case when the mode on the curved waveguide is not a small perturbation of a mode on the straight guide, but is guided essentially by the outer boundary (modes of this nature are known as 'edge-guided' modes).

Unlike the slab, it is apparently impossible to obtain an exact eigenvalue equation for the curved fiber. An approximate field analysis is complicated by the fact that field matching at the dielectric boundary requires a knowledge of the functional form taken there by the radiation condition. At infinity, this demands outgoing or attenuated fields, but, as shown by Lewin[34] in 1974, this requirement takes a different form locally. The appropriate field structure for both the slab and the cylinder was deduced and the results applied to the bent optical fiber. Arnaud,[35] also in 1974, imagined the curved fiber to be surrounded by a metallic cylinder which supported whispering gallery modes, and evaluated the radiation as coupling to these modes, in the limit as the metallic cylinder was removed to infinity. In both these papers, the external fields are expressed as Fourier integrals with respect to the direction normal to the plane of the bend, which has the effect of reducing the local form of the radiation condition to that for a slab, with a spectral variable appearing as a parameter. Using this representation, Chang and Kuester[36] in 1976 obtained an expression for radiation loss of dielectric guides of arbitrary cross-section, based upon a perturbation procedure. Further work along these lines has been done by Marcuse[37] in 1976, who deals with the case when the internal fields of the fiber can no longer be considered as small perturbations of those of the straight fiber, similar to edge-guided modes on a slab.

Snyder, White and Mitchell[38] in 1975 obtained the same radiation loss formula for a curved fiber by using a Poynting-vector approach to the far-field radiation, assumed to be caused by equivalent surface currents on the fiber. This same approach has been used by Maley[39] in 1974 for the case of scattering from a bent dielectric slab between two straight sections, and he shows that a power loss inversely proportional to the square of the radius of curvature is introduced by the junction discontinuities, as is the case for the analogous closed waveguide problem. Since the continuous radiation due to the curvature itself (rather than the change in curvature) decreases exponentially with the radius of curvature, the junction scattering dominates for very small curvatures. The method was extended by Maley[40] in 1976 to cover the optical fiber case. The mechanical deformation effect of the fiber bending on the optical properties was considered by Hannay,[41] and the effect of a parabolic refractive index on the design of bends in multimode fibers was considered by Sawa and Kumagai,[42] both in 1976. The *increase* in radiation losses due to curvature for slightly leaky modes on hollow dielectric or imperfect metallic waveguides, a sort of composite of open and closed waveguide behavior, was investigated by Marcatili and Schmeltzer[43] in 1964.

An alternative approach to curvature radiation problems is through a change of variables, which enables the wave equation in curvilinear co-ordinates to be expressed in a form equivalent to that for the straight structure. The curvature effects then appear in a way reminiscent of a variable refractive index and suitable techniques exist for obtaining solutions. Sveshnikov's paper[16] is an example in the closed guide case, but the method itself is much older, having been used to obtain the 'flat-earth' equivalence for both curvature and atmospheric refractive index variations. The method was applied by Felsen and Maurer,[21] in their 1970 review paper, to a curved dielectric slab on a metal plate, in 1973 by Chang and Barnes[31] to radiation from an asymmetric dielectric slab, and also in 1975 to more general slab structures by Heiblum and Harris.[32] The method provides a certain amount of physical insight into the nature of the local form of the radiation condition, since it predicts a caustic in the surface wave above a convex surface at which the equivalent wave vector in the modified co-ordinate system changes character and leads to an imperfect reflection and hence an incoming local attenuated wave. Above a *concave* surface the caustic is perfectly reflecting, and the reflection there is associated with whispering-gallery modes. All these features enter in the very complex problem of the exact calculation of the radiation from the junction between straight and curved open structures, and at the time of writing have not been completely resolved.

The present monograph has been designed to bring out the salient features of much of the above material, concentrating mainly on asymptotic and perturbation techniques. The first two chapters deal with Tang's orthogonalization procedures and with the bi-complex variable techniques of decoupling the vector wave equation. Chapter 3 introduces the twisted co-ordinate system with application to rectangular waveguides. The next three chapters are concerned with curved guides, both rectangular and circular. Calculations are performed for the propagation coefficient, reflection, mode-conversion, mode-coupling and eigenfunctions for a variety of configurations by the asymptotic method. Chapter 7 discusses the radiation condition at curved surfaces from a number of points of view, and results are applied to dielectric slabs and fibers in the subsequent two chapters. The last chapters attempt a consideration of the very complex problem of the discontinuity at a junction of a straight and curved radiating structure, and indicate the latest results available at the time of writing. Several appendices discuss detailed material that would probably have been somewhat distracting if included in their respective chapters. And, rather specially, because it has nothing to do with the main topic of this book, the last appendix gives an analysis of the resonant window in rectangular waveguide. The reason for its inclusion is that this material, first published in 1951, has long since been out of print; but continued requests for copies indicated that it is probably of value to re-issue it, and the present volume appeared to be a not unsuitable vehicle.

References

1 J.R. Wait: 'Electromagnetic waves in stratified media' (Oxford, Pergamon Press, 1970)

2 H.C. Pocklington: 'Electric oscillations on wires', *Proc. Cambridge Phil. Soc.*, vol. 9, part 7, 1897, pp. 324–332

3 H. Buchholz: 'Der Einfluss der Krümmung von rechteckigen Hohleitern auf das Phasenmass ultrakurzer Wellen', *E.N.T.*, vol. 16, 1939, pp. 73–85

4 R.E. Marshak: 'Theory of circular bends in rectangular wave guides', Radiation Laboratory Report 43–45, June 24, 1943

5 M. Jouguet: 'Propagation dans les tujaux courbés', *Comptes Rendus – Académie des Sciences,* Paris, Feb 18, 1946, March 4, 1946 and Jan. 6, 1947

6 M. Jouget: 'Les effets de la courbure et des discontinuités de courbure sur la propagation des ondes dans les guides à section rectangulaire', *Annales des Télécommuncation,* t.2, no. 3, 1947, pp. 78–94

7 M. Jouguet: 'Effects of curvature on the propagation of electromagnetic waves in guides of circular cross-section', *Câbles et Transmission,* Paris, vol. 1, no. 2, July 1947, pp. 133–153

8 K. Riess: 'Electromagnetic waves in a bent pipe of rectangular cross-section', *Q. Appl. Math.*, vol. 1, 1944, pp. 328–333

9 S. A. Schelkunoff: Note on the above paper, *Q. Appl. Math.*, vol. 2, 1944, p. 171

10 S.O. Rice: 'Reflections from circular bends in rectangular wave guides – matrix theory', *B.S.T.J.*, vol. 27, 1948, pp. 305–349

11 S.A. Schelkunoff: 'Conversion of Maxwell's equations into generalized telegraphist's equations', *B.S.T.J.*, vol. 34, 1955, pp. 995–1043

12 W.J. Albersheim: 'Propagation of TE_{01} waves in curved waveguides', *B.S.T.J.*, Jan. 1949, vol. 28, pp. 1–32

13 W. Sollfrey: 'Wave propagation on helical wires', *J. Appl. Phys.*, vol. 22, no. 7, July 1951, pp. 905–910

14 L. Lewin: 'Propagation in curved and twisted waveguides of rectangular cross-section', *Proc. IEE,* vol. 102, part B, no. 1, Jan. 1955, pp. 75–80

15 R.A. Waldron: 'Theory of the helical waveguide of rectangular cross-section', *J. Brit. IRE,* Oct. 1957, pp. 577–592

16 A.G. Sveshnikov: 'On the propagation of radiowaves in gently curved guides', *Radiotekhnika i Electronika,* vol. 1, 1956, pp. 1222–1229

17 S.P. Morgan: 'Theory of curved circular waveguide containing an inhomogeneous dielectric', *B.S.T.J.*, Sept. 1957, vol. 36, pp. 1209–1251

18 J.P. Quine: 'E and H-plane bends for high-power oversized rectangular waveguide', *IEEE Trans. MTT,* vol. 13, Jan. 1965, pp. 54–63

19 J.A. Cochran and R.G. Pecina: 'Mode propagation in continuously curved waveguides', *Radio Science,* vol. 1, no. 6, June 1966, pp. 679–696

20 E. Burstein and L. Solov'ev: 'On the propagation of the fundamental wave between two parallel surfaces', *Dokl. Akad. Nauk. SSSR,* vol. 101, 1955, pp. 465–468

21 S.J. Maurer and L.B. Felsen: 'Ray methods for trapped and slightly leaky modes in multilayered or multiwave regions', *IEEE Trans. MTT,* vol. 18, Sept. 1970, pp. 584–595

22 D.S. Ahluwalia, J.B. Keller and B.J. Matkowsky: 'Asymptotic theory of propagation in curved and non-uniform waveguides', *J. Acoustic. Soc. Am.,* vol. 55, no. 1, Jan. 1974, pp. 7–12

23 E. Bahar: 'Fields in waveguide bends expressed in terms of coupled local annular waveguide modes', *IEEE Trans. MTT,* vol. 17, April 1969, pp. 210–217

24 E. Bahar and G. Govindarajan: 'Rectangular and annular modal analyses of multimode waveguide bends', *IEEE Trans. MTT,* vol. 21, Dec. 1973, pp. 819–824

25 C.P. Bates: 'Intermodal coupling at the junction between straight and curved waveguides', *B.S.T.J.,* vol. 48, Sept. 1969, pp. 2259–2280

26 C.H. Tang: 'An orthogonal coordinate system for curved pipes', *IEEE Trans. MTT,* vol. 18, Jan. 1970, p. 69

27 L. Lewin: 'A decoupled formulation of the vector wave equation in orthogonal curvilinear coordinates, with application to ferrite-filled and curved waveguides of general cross-section, *IEEE Trans. MTT,* vol. 20, no. 5, May 1972, pp. 338–342

28 R.D. Richtmyer: 'Dielectric resonators', *J. Appl. Phys.,* vol. 10, June 1939, pp. 391–398

29 E.A.J. Marcatili: 'Bends in optical dielectric guides', *Bell Syst. Tech. J.,* vol. 48, Sept. 1969, pp. 2103–2132

30 D. Marcuse: 'Bending losses of the asymmetric slab waveguide', *Bell. Syst. Tech. J.,* vol. 50, 1971, pp. 2551–2563

31 D.C. Chang and F.S. Barnes: 'Reduction of radiation loss in a curved dielectric slab waveguide', Sci. Rept. no. 2 (AFOSR-72-2417), Dept. of Elec. Eng., Univ. of Colo., Boulder, Colo. 80309 (July 1973)

32 M. Heiblum and J.H. Harris: 'Analysis of curved optical waveguides by conformal transformation', *IEEE J. Quantum Electron.,* vol. 11, Feb. 1975, pp. 75–83. Correction, *ibid.,* vol. 12, p. 313, May 1976

33 S. Kawakami, M. Miyagi and S. Nishida: 'Bending losses of dielectric slab optical waveguide with double or multiple claddings', *Appl. Optics,* vol. 14, Nov. 1975, pp. 2588–2597. Correction, *ibid.,* vol. 15, p. 1681, July 1976

34 L. Lewin: 'Radiation from curved dielectric slabs and fibers', IEEE Trans. MTT, vol. 22, no. 7, July 1974, pp. 718–727. Correction, *ibid.,* vol. 23, p. 779, 1975

35 J.A. Arnaud: 'Transverse coupling in fiber optics – Part III: Bending losses', *B.S.T.J.,* vol. 53, Sept. 1974, pp. 1379–1394

36 D.C. Chang and E.F. Kuester: 'Radiation and propagation of a surface-wave mode on a curved open waveguide of arbitrary cross section', *Radio Science* vol. 11, May 1976, pp. 449–457

37 D. Marcuse: 'Field deformation and loss caused by curvature of optical fibers', *J. Opt. Soc. Amer.,* vol. 66, April 1976, pp. 311–320

38 A.W. Snyder, I. White and D.J. Mitchell: 'Radiation from bent optical waveguides', *Electronics Letters,* July 24, 1975, vol. 11, no. 15, pp. 332–333
39 S.W. Maley: 'Radiation from a circular bend between two discontinuities in a dielectric slab waveguide', Sci. Rept. no. 10, Dept. of Elec. Eng. University of Colo., Boulder, Colo. 80309 (Dec. 1974)
40 S.W. Maley: 'Radiation from a circular bend between two discontinuities in dielectric slab waveguides and in circular cylindrical dielectric waveguides'. Sci. Rept. no. 18, Dept. of Elec. Eng., Univ. of Colo., Boulder, Colo. 80309 (June 1976)
41 J.H. Hannay: 'Mode coupling in an electrically deformed optical fiber', *Electronics Letters,* Apr. 1, 1976, vol. 12, no. 7, pp. 173–174
42 S. Sawa and N. Kumagai: 'Analysis of optical waveguide consisting of a square-law lenslike medium, and its analogies to circular TE_{01} waveguide', *IEEE Trans. MTT,* vol. 24, no. 7, July 1976, pp. 441–455
43 E.A.J. Marcatili and R.A. Schmeltzer: 'Hollow metallic and dielectric waveguides for long distance optical transmission and lasers', *Bell Syst. Tech. J.,* vol. 43, 1964, pp. 1783–1809

Chapter 1

Orthogonal co-ordinate systems for curved structures

1.1 Maxwell's equations

In this book we shall be concerned with time-harmonic solutions of Maxwell's equations in source-free linear isotropic media. The equations accordingly reduce to the simpler forms

$$\text{curl}\, H = j\omega\epsilon E \tag{1.1}$$

$$\text{curl}\, E = -j\omega\mu H \tag{1.2}$$

where

E is the electric fieldstrength
H is the magnetic fieldstrength
ϵ is the (complex) permittivity
μ is the permeability
$j = \sqrt{(-1)}$ is used in the combination $e^{j\omega t}$ to represent the time-harmonic variation.

In this context j will be restricted to this application only, i.e. in conjunction with ωt, with the factor $e^{j\omega t}$ normally omitted. Later we shall use bi-complex forms with i as the basis for a complex field system having no relation to time variations. It is clear that (1.1) and (1.2) are completely general vector differential relations independent of any co-ordinate system. The usual div relations are implicit, since div curl $\equiv 0$, and there is no free charge in a source-free region.

Useful solutions of (1.1) and (1.2) can be obtained in only a few orthogonal co-ordinate systems. Perturbations of such systems normally introduce non-orthogonal features that further complicate the construction of solutions. We shall be working with one such non-orthogonal system in relation to the twisted rectangular waveguide, but generally it is preferable to work with an orthogonal system where possible. A means whereby a suitable orthogonal system can be constructed from a more general one has been developed by Tang[1] in relation to applications to waveguides whose axes are curved or twisted in space; it is discussed in the following sections.

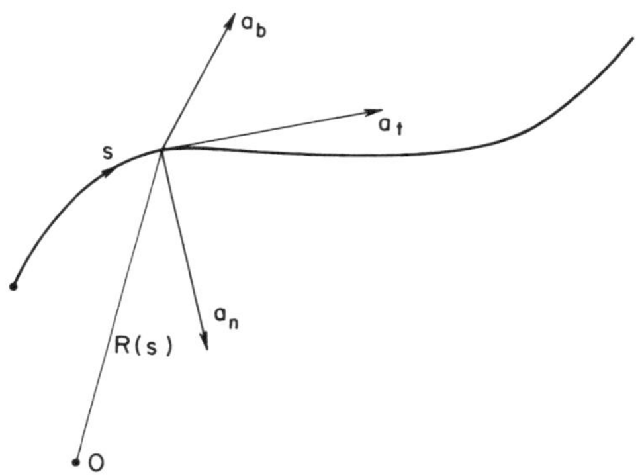

Fig. 1.1 Serret-Frenet frame for a curve in space

1.2 Curvature relations

Fig. 1.1 shows a smooth curve in space described by a position vector $\boldsymbol{R}(s)$, a function of s, the arc length measured from an arbitrary point. At any point we can set up a unit vector $\boldsymbol{a}_t$ in the direction of the tangent at that point. Clearly we have the vector relation $\boldsymbol{R}(s) + \boldsymbol{a}_t ds = \boldsymbol{R}(s + ds)$, from which we deduce

$$\boldsymbol{a}_t = \boldsymbol{R}' \tag{1.3}$$

the prime representing a differentiation with respect to s.

In going from s to $s + ds$, this unit tangent vector rotates by an angle $d\theta$, and the rotation defines a plane in which the principal normal to the curve is located. If $\boldsymbol{a}_n$ is the unit vector in the direction of the principal normal, then $\boldsymbol{a}_n d\theta$ is the third side of the triangle whose other two sides are $\boldsymbol{a}_t(s)$ and $\boldsymbol{a}_t(s + ds)$. Hence $\boldsymbol{a}_t(s + ds) = \boldsymbol{a}_t(s) + \boldsymbol{a}_n d\theta$, from which $\boldsymbol{a}_n = \boldsymbol{a}_t'(ds/d\theta)$. Now the curvature χ is defined as the rate of rotation of the tangent with arc length, or $\chi = d\theta/ds$. From this we get a relation between the unit normal, unit tangent and the curvature:

$$\boldsymbol{a}_n = \boldsymbol{a}_t'/\chi \tag{1.4}$$

Since $\boldsymbol{a}_t' = \boldsymbol{R}''$ (1.4) gives, on taking the magnitude of both sides, the curvature in terms of the second derivative of the position vector,

$$\chi^2 = (\boldsymbol{R}''\cdot\boldsymbol{R}'') \tag{1.5}$$

A right-handed rotation from the tangent to the principal normal defines the direction of the binormal. Hence the unit vector $\boldsymbol{a}_b$ in that direction is given by $\boldsymbol{a}_b = \boldsymbol{a}_t \times \boldsymbol{a}_n$, or

$$\boldsymbol{a}_b = \boldsymbol{R}' \times \boldsymbol{R}''/\chi \tag{1.6}$$

An additional relation involving $\boldsymbol{a}_b$ relates these unit vectors to the torsion τ. As the point s moves to $s+ds$ the principal normal $\boldsymbol{a}_n(s)$ becomes $\boldsymbol{a}_n(s+ds)$. The difference $\boldsymbol{a}_n(s+ds) - \boldsymbol{a}_n(s) = \boldsymbol{a}_n' ds$ is perpendicular to $\boldsymbol{a}_n$ (since the two vectors are of unit length and form neighboring radii of a unit circle). Hence $\boldsymbol{a}_n'$, which can be expressed in terms of multiples of $\boldsymbol{a}_n$, $\boldsymbol{a}_b$ and $\boldsymbol{a}_t$, will involve only the latter two, so that we can write

$$\boldsymbol{a}_n' = A\boldsymbol{a}_b + B\boldsymbol{a}_t \tag{1.7}$$

By the definition of torsion the coefficient $A = \tau$.

Substituting into (1.7) from (1.3), (1.4) and (1.6) gives

$$\boldsymbol{R}'''/\chi - \boldsymbol{R}''\chi'/\chi^2 = A\boldsymbol{R}' \times \boldsymbol{R}''/\chi + B\boldsymbol{R}' \tag{1.8}$$

Now $\boldsymbol{R}' \times \boldsymbol{R}''/\chi = \boldsymbol{a}_b$ and is perpendicular to $\boldsymbol{a}_t = \boldsymbol{R}'$ and $\boldsymbol{a}_n = \boldsymbol{R}''/\chi$. Hence, if we take the scalar product of (1.8) with $\boldsymbol{R}' \times \boldsymbol{R}''$ we isolate the coefficient A to give

$$A = \tau = (\boldsymbol{R}' \times \boldsymbol{R}'')\cdot\boldsymbol{R}'''/(\boldsymbol{R}''\cdot\boldsymbol{R}'') \tag{1.9}$$

This relation gives the torsion in terms of derivatives of the position vector. From $\boldsymbol{a}_t \cdot \boldsymbol{a}_n = 0$ we have $\boldsymbol{R}'\cdot\boldsymbol{R}'' = 0$, and differentiating gives $\boldsymbol{R}''\cdot\boldsymbol{R}'' = -\boldsymbol{R}'\cdot\boldsymbol{R}'''$. Taking the scalar product of (1.8) with $\boldsymbol{a}_t = \boldsymbol{R}'$ accordingly gives $B = \boldsymbol{R}'\cdot\boldsymbol{R}'''/\chi = -(\boldsymbol{R}''\cdot\boldsymbol{R}'')/\chi = -\chi$ from (1.5).

Substituting in (1.7) enables it to be put in the form

$$\boldsymbol{a}_b = (\boldsymbol{a}_n' + \chi\boldsymbol{a}_t)/\tau \tag{1.10}$$

Differentiating the relation $\boldsymbol{a}_b = \boldsymbol{a}_t \times \boldsymbol{a}_n$ gives $\boldsymbol{a}_b' = \boldsymbol{a}_t' \times \boldsymbol{a}_n + \boldsymbol{a}_t \times \boldsymbol{a}_n'$. From (1.4) $\boldsymbol{a}_t' = \chi\boldsymbol{a}_n$ so $\boldsymbol{a}_t' \times \boldsymbol{a}_n = 0$; and from (1.10), $\boldsymbol{a}_t \times \boldsymbol{a}_n' = \tau\boldsymbol{a}_t \times \boldsymbol{a}_b = -\tau\boldsymbol{a}_n$. Hence

$$\boldsymbol{a}_b' = -\tau\boldsymbol{a}_n \tag{1.11}$$

1.3 The Serret-Frenet frame and the Darboux vector

The three mutually orthogonal vectors $\boldsymbol{a}_t$, $\boldsymbol{a}_b$ and $\boldsymbol{a}_n$ define a right-handed orthogonal co-ordinate system at any point of the curve, and is known as the Serret-Frenet frame. As the point s moves along the curve the frame rotates. If the curve is a plane curve the rotation is about the axis $\boldsymbol{a}_b$; the effect of the torsion is to provide additional rotation about $\boldsymbol{a}_t$.

A vector which measures the rate of rotation of the frame is the Darboux vector $\boldsymbol{\delta}$, defined by

$$\boldsymbol{\delta} = \tau \boldsymbol{a}_t + \chi \boldsymbol{a}_b \tag{1.12}$$

Equations (1.4), (1.7) and (1.11) can be put in the form

$$\begin{aligned} \boldsymbol{a}_t' &= \chi \boldsymbol{a}_n = \boldsymbol{\delta} \times \boldsymbol{a}_t \\ \boldsymbol{a}_n' &= -\chi \boldsymbol{a}_t + \tau \boldsymbol{a}_b = \boldsymbol{\delta} \times \boldsymbol{a}_n \\ \boldsymbol{a}_b' &= -\tau \boldsymbol{a}_n = \boldsymbol{\delta} \times \boldsymbol{a}_b \end{aligned} \tag{1.13}$$

From (1.13) the Darboux vector can be interpreted as an angular velocity vector describing the rate of rotation of the trihedral ($\boldsymbol{a}_t$, $\boldsymbol{a}_n$, $\boldsymbol{a}_b$) as it moves along the curve.

1.4 Derived co-ordinate system

The Serret-Frenet frame can be used as the basis for setting up an orthogonal co-ordinate system in the neighbohood of the curve. Actually on the curve the Serret-Frenet frame itself constitutes just such an orthogonal system, but as we shall show, for points off the curve, the effect of the torsion is to destroy the orthogonality: the further from the curve the more oblique does the co-ordinate system become.

Let r, in Fig. 1.2, be the position vector of an arbitrary point P in the vicinity of the curve, and let P_0 be a point on the curve corresponding to P. The nature of the correspondence is such that the plane through P_0 perpendicular to the curve contains P, so that $r = R + \boldsymbol{\rho}$ where $\boldsymbol{\rho}$ lies in the transverse plane determined by $\boldsymbol{a}_n$ and $\boldsymbol{a}_b$ at P_0. Accordingly, we can put $\boldsymbol{\rho} = n\boldsymbol{a}_n + b\boldsymbol{a}_b$ where n and b are co-ordinate distances in the (local) directions of $\boldsymbol{a}_n$ and $\boldsymbol{a}_b$ respectively. The point P_0 determines a value of s corresponding to P, and P can be specified by the values of (n, b, s). This trio can therefore be used as co-ordinates of P, with n and b playing the roles of transverse co-ordinates and s that of axial co-ordinate.

From $\boldsymbol{\rho} = n\boldsymbol{a}_n + b\boldsymbol{a}_b$ we get $d\boldsymbol{\rho} = dn\boldsymbol{a}_n + db\boldsymbol{a}_b + (n\boldsymbol{a}_n' + b\boldsymbol{a}_b')ds = dn\boldsymbol{a}_n + db\boldsymbol{a}_b + \boldsymbol{\delta} \times (n\boldsymbol{a}_n + b\boldsymbol{a}_b)ds$ on using (1.13). Hence the differential $d\boldsymbol{r}$ can be expressed in this co-ordinate system by

$$d\boldsymbol{r} = \boldsymbol{a}_n dn + \boldsymbol{a}_b db + [\boldsymbol{a}_t + \boldsymbol{\delta} \times (n\boldsymbol{a}_n + b\boldsymbol{a}_b)]\, ds \qquad (1.14)$$

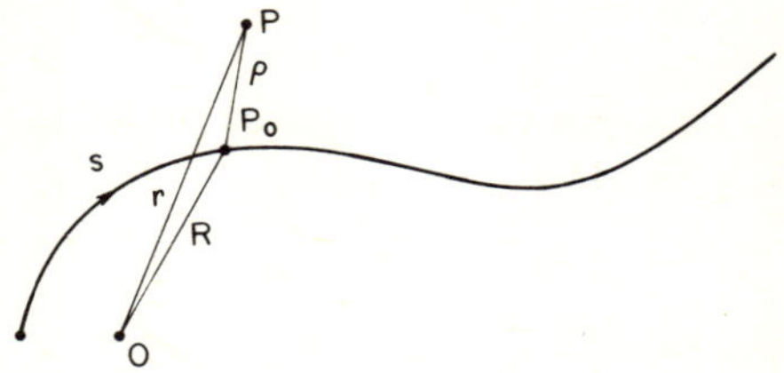

Fig. 1.2 Construction of local co-ordinate system

From this, by taking the magnitudes of both sides, we get the line element

$$(dr)^2 = (dn)^2 + (db)^2 + [(1 - \chi n)^2 + \tau^2(n^2 + b^2)]\,(ds)^2 + 2\tau(ndb - bdn)ds \qquad (1.15)$$

Except on the curve, where $n = b = 0$, (1.15) describes an oblique co-ordinate system owing to the presence of the last term in the torsion. Apart from the case of a plane curve the Serret-Frenet frame does not, of itself, provide an orthogonal co-ordinate system.

1.5 An orthogonal co-ordinate system

Since it is the torsion that destroys the orthogonality, an orthogonal system can be devised by producing an alternative frame with zero rotation around $\boldsymbol{a}_t$, i.e. a frame with a rotation rate $-\tau$ with respect to the Serret-Frenet frame. Denote the new frame by unit vectors $\boldsymbol{a}_\nu$, $\boldsymbol{a}_\beta$, $\boldsymbol{a}_t$, and let the angle between $\boldsymbol{a}_n$ and $\boldsymbol{a}_\nu$ be ψ, a function of s such that

$$\psi'(s) = -\tau(s) \qquad (1.16)$$

Then for the frame $\boldsymbol{a}_\nu$, $\boldsymbol{a}_\beta$, $\boldsymbol{a}_t$ the Darboux vector becomes, since there is now no net torsion,

$$\boldsymbol{\delta}_0 = \chi\boldsymbol{a}_b = \chi(\boldsymbol{a}_\nu \sin\psi + \boldsymbol{a}_\beta \cos\psi) \qquad (1.17)$$

Since $n = \nu \cos \psi - \beta \sin \psi$ the expression for the line element in the (ν, β, s) system reduces to

$$(dr)^2 = (d\nu)^2 + (d\beta)^2 + [1 - \chi(\nu \cos \psi - \beta \sin \psi)]^2 (ds)^2 \quad (1.18)$$

This form verifies the mutual orthogonality of ν, β and s, and identifies the metric coefficients as

$$h_\nu = h_\beta = 1, \quad h_s = [1 - \chi(\nu \cos \psi - \beta \sin \psi)] \quad (1.19)$$

Accordingly, Maxwell's equations in orthogonal curvilinear form can now be constructed in the (ν, β, s) system.

1.6 Waveguide applications

The preceding results apply to the case of a rectangular waveguide whose axis follows the curve $\boldsymbol{R}(s)$. The cross-section can be defined by the relations $\nu = \pm a/2$ and $\beta = \pm b/2$ where a and b are the sides of the cross-section.

For shapes of cross-section other than rectangular, consider the transformation

$$\nu = \nu(u, v), \quad \beta = \beta(u, v) \quad (1.20)$$

to orthogonal curvilinear co-ordinates u, v.

The orthogonality condition on u and v takes the form

$$\frac{\partial \nu}{\partial u}\frac{\partial \nu}{\partial v} + \frac{\partial \beta}{\partial u}\frac{\partial \beta}{\partial v} = 0 \quad (1.21)$$

The metric coefficients in the new co-ordinate system are

$$\begin{aligned} h_u &= [(\partial \nu/\partial u)^2 + (\partial \beta/\partial u)^2]^{1/2} \\ h_v &= [(\partial \nu/\partial v)^2 + (\partial \beta/\partial v)^2]^{1/2} \end{aligned} \quad (1.22)$$

with h_s as in (1.19).

For example, by changing from (ν, β) to polar co-ordinates (ρ, ϕ) a curved circular pipe is defined by $\rho^2 = \nu^2 + \beta^2 = \rho_0^2$, say. The corresponding metric coefficients become

$$h_\rho = 1, \quad h_\phi = \rho, \quad h_s = 1 - \chi\rho \cos(\phi + \psi) \quad (1.23)$$

Suppose the guide axis follows a helix of radius R_0 and period L. It is readily verified that $\boldsymbol{R}(s) = \boldsymbol{a}_x \cos \theta + \boldsymbol{a}_y \sin \theta + z\boldsymbol{a}_z$ with $s = z[1 + (2\pi R_0/L)^2]^{1/2}$, $\theta = 2\pi z/L$. The differentiations required by (1.5) and (1.9) are readily performed and give

$$\chi = \frac{-1/R_0}{1 + (L/2\pi R_0)^2}$$
$$\tau = \frac{2\pi L}{L^2 + (2\pi R_0)^2} \qquad (1.24)$$

From (1.16), ψ in (1.23) becomes $-2\pi Ls/[L^2 + (2\pi R_0)^2]$. If the guide is bent in one plane the helix degenerates to a circle of radius R_0. Then $L = 0$, the torsion vanishes, and the metric coefficients reduce to

$$h_\rho = 1, \quad h_\phi = \rho, \quad h_s = 1 + (\rho/R_0)\cos\phi \qquad (1.25)$$

In this formula the direction of the center of curvature of the guide axis is at $\phi = \pi$.

If the transformation of (1.20) is conformal then the orthogonality condition (1.21) is automatically satisfied, and then $h_u = h_v$. It may be noted that the co-ordinate system thus obtained differs from a conventional cylindrical co-ordinate system with a constant direction axis only by the axial metric coefficient h_s, which is given by (1.19) instead of being unity. Of course, in accordance with (1.16), the transverse co-ordinates, and with them the guide cross-section, twist in order to compensate the effects of the torsion. For a straight guide the curvature is zero, and the two systems then coincide, as of course they must do.

Reference

1 Tang, C.H.: 'An orthogonal coordinate system for curved pipes', *IEEE Trans. MTT-18*, p. 69, Jan. 1970

Chapter 2

Maxwell's equations for curved waveguides

2.1 Rotating polarization in straight waveguide

The example of rotating polarization will be used to illustrate the method of combining transverse field components. As applied to a cylindrical waveguide with straight axis the equation curl curl $E - k^2 E = 0$ reduces to the following three equations:

$$\begin{aligned} \nabla^2 E_\rho + k^2 E_\rho - E_\rho/\rho^2 &= (2/\rho^2)\,\partial E_\phi/\partial\phi \\ \nabla^2 E_\phi + k^2 E_\phi - E_\phi/\rho^2 &= (2/\rho^2)\,\partial E_\rho/\partial\rho \\ \nabla^2 E_z + k^2 E_z &= 0 \end{aligned} \tag{2.1}$$

where $\nabla^2 \equiv$ div grad is the Laplacian operating on the ρ, ϕ, and z scalar components of the electric field. It is the appearance of the terms on the right of the first two equations that is typical of the form taken by the wave equation in curvilinear co-ordinates. These terms couple the equations together, complicating the solution.

Endean and Allen[1] have shown that by forming the complex combinations

$$\begin{aligned} E_1 &= E_\rho + jE_\phi \\ E_2 &= E_\rho - jE_\phi \end{aligned} \tag{2.2}$$

the first two equations of (2.1) can be combined to yield the *decoupled* equations

$$\begin{aligned} \nabla^2 E_1 + (2j/\rho^2)\,\partial E_1/\partial\phi - E_1/\rho^2 + k^2 E_1 &= 0 \\ \nabla^2 E_2 - (2j/\rho^2)\,\partial E_2/\partial\phi - E_2/\rho^2 + k^2 E_2 &= 0 \end{aligned} \tag{2.3}$$

These can now be solved by standard techniques. If the field components vary as $e^{jm\phi}$ the terms not involving ρ derivatives become $[k^2 - (m+1)^2/\rho^2]\,E_1$ and $[k^2 - (m-1)^2/\rho^2]\,E_2$ and clearly lead to solutions involving Bessel functions of order $m \pm 1$. As the ϕ variation appears in the final solution in the combination $e^{j(\omega t + m\phi)}$ the result is obviously useful for expressing solutions involving rotating fields. However, the method is more powerful than this. The more general features are discussed in the next section.

2.2 Bi-complex forms

Since the equations in (2.1) are real in form it is clear that the use of j in (2.2) was purely as a convenience to achieve the decoupled pair (2.3). It really need have nothing to do with time variations. In fact the fields could have been expressed in a form showing a cos ωt variation explicitly and the decoupled equations (2.3) would still have followed. In this case E_1 and E_2 would have been complex conjugates with E_ρ and E_ϕ found by taking the real and imaginary parts. In order to retain the convenience of using $e^{j\omega t}$ to represent the time variation it is necessary to use a different imaginary symbol to achieve the decoupling. We shall here use i and define

$$E_1 = E_\rho + iE_\phi$$

$$E_2 = E_\rho - iE_\phi \tag{2.4}$$

Clearly E_1 and E_2 are complex conjugates (in the field of i) irrespective of how they may vary with j. In this bi-complex system,[2] we have $i^2 = -1$, $j^2 = -1$ but cannot take $ij = -1$. As far as i is concerned j behaves just like any ordinary algebraic symbol, and operations like Re (taking the real part of) need to specify whether it is over the field of i or j. In the rest of this book we shall not have occasion to need to extract the real time form from $e^{j\omega t}$, so it will be understood henceforth that the operations Re and Im (imaginary part of) refer only to terms in the field of i.

By means of (2.4) the equations for E_1 and E_2 can be written

$$\nabla^2 E \pm (2i/\rho^2)\,\partial E/\partial\phi - E/\rho^2 + k^2 E = 0 \tag{2.5}$$

where E stands for E_1 when the + sign is taken, and for E_2 with the − sign. Although (2.5) is really two equations it can be treated as a single one in which the quantity i could have either sign. By this means the equations are *compacted*. A more general use of this method to curved guides of arbitrary cross-section follows.

2.3 Decoupled vector wave equation

The formulas of Chapter 1, especially those of Section 1.6, give the co-ordinate metric coefficients for a waveguide of arbitrary but constant cross-section following a specified axial path. We shall here use u_3 to represent the axial co-ordinate and u_1 and u_2 for the transverse co-ordinates. These latter, as well as the metric coefficients h_1 and h_2, are the same as they would have been in a straight waveguide, and are therefore independent of u_3. The metric coefficient h_3, however, is a function of u_1 and u_2, and may also be a function of u_3. In this co-ordinate system Maxwell's equations become

$$\left.\begin{aligned}
\frac{\partial(h_3E_3)}{\partial u_2} - \frac{\partial(h_2E_2)}{\partial u_3} &= -j\omega\mu h_2h_3H_1 \\
\frac{\partial(h_1E_1)}{\partial u_3} - \frac{\partial(h_3E_3)}{\partial u_1} &= -j\omega\mu h_3h_1H_2 \\
\frac{\partial(h_2E_2)}{\partial u_1} - \frac{\partial(h_1E_1)}{\partial u_2} &= -j\omega\mu h_1h_2H_3
\end{aligned}\right\} \tag{2.6}$$

$$\left.\begin{aligned}
\frac{\partial(h_3H_3)}{\partial u_2} - \frac{\partial(h_2H_2)}{\partial u_3} &= j\omega\epsilon h_2h_3E_1 \\
\frac{\partial(h_1H_1)}{\partial u_3} - \frac{\partial(h_3H_3)}{\partial u_1} &= j\omega\epsilon h_3h_1E_2 \\
\frac{\partial(h_2H_2)}{\partial u_1} - \frac{\partial(h_1H_1)}{\partial u_2} &= j\omega\epsilon h_1h_2E_3
\end{aligned}\right\} \tag{2.7}$$

These equations are the particular case of a more general formulation[3] involving an axially magnetized medium in the guide.

It is convenient to define, for the axial fields, the combinations

$$E = h_3E_3, \quad H = h_3H_3 \tag{2.8}$$

For the transverse components define the combinations

$$E_t = E_1 + iE_2, \quad H_t = H_1 + iH_2 \tag{2.9}$$

Since E_1 and E_2 are both real (in i) they are directly related to the real and imaginary parts of E_t: and similarly for H_t.

We define two transverse differential operators which are complex conjugates of each other:

$$D, D^* \equiv \frac{1}{h_1}\frac{\partial}{\partial u_1} \pm i\frac{1}{h_2}\frac{\partial}{\partial u_2} \tag{2.10}$$

Noting that h_1 and h_2 have been taken independent of u_3 it can be readily verified that the first two equations of (2.6) can be combined into the form

$$DE = \partial E_t/\partial u_3 - ij\omega\mu h_3H_t \tag{2.11}$$

Similarly the first two equations of (2.7) become

$$DH = \partial H_t/\partial u_3 + ij\omega\epsilon h_3E_t \tag{2.12}$$

To deal with the third equations of (2.6) and (2.7) we need a further operator $\bar{D}$ given by

$$\bar{D} \equiv \frac{1}{h_1 h_2}\left[\frac{\partial}{\partial u_2}(h_1 \qquad) + i\frac{\partial}{\partial u_1}(h_2 \qquad)\right] \tag{2.13}$$

and the equations become

$$h_3 \operatorname{Re}(\bar{D}E_t) = j\omega\mu H \tag{2.14}$$

$$h_3 \operatorname{Re}(\bar{D}H_t) = -j\omega\epsilon E \tag{2.15}$$

As explained earlier, Re and Im throughout refer to the field of complex quantities in i.

The shortcoming of (2.14) and (2.15) for further analysis is the presence of the Re operator. It can, in effect, be removed if we can also obtain $ih_3 \operatorname{Im}(\bar{D}E_t)$ and $ih_3 \operatorname{Im}(\bar{D}H_t)$ and add these to the equations. To do this we first consider the double operator $\bar{D}D$, which can be readily verified to reduce to

$$\bar{D}D = \frac{i}{h_1 h_2}\left\{\frac{\partial}{\partial u_1}\left(\frac{h_2}{h_1}\frac{\partial}{\partial u_1} \qquad\right) + \frac{\partial}{\partial u_2}\left(\frac{h_1}{h_2}\frac{\partial}{\partial u_2} \qquad\right)\right\} \tag{2.16}$$

Clearly the result is imaginary if the operand is real, as are both E and H. Operating on (2.11) with $\bar{D}$ and taking the real part gives

$$0 = \operatorname{Re}\bar{D}\,[\partial E_t/\partial u_3 - ij\omega\mu h_3 H_t] \tag{2.17}$$

Now $\partial/\partial u_3$ commutes with $\bar{D}$ since h_1 and h_2 are independent of u_3; (2.17) becomes, on using (2.14) for $\operatorname{Re}(\bar{D}E_t)$:

$$0 = \frac{\partial(j\omega\mu H/h_3)}{\partial u_3} + \operatorname{Im} j\omega\mu\bar{D}(h_3 H_t) \tag{2.18}$$

It is readily verified that, if x and y are any two variables, the effect of $\bar{D}$ operating on the product xy can be expressed as

$$\bar{D}(xy) = x\bar{D}(y) + iyD^*(x) \tag{2.19}$$

Applying this to (2.18) with $x = h_3$, $y = H_t$ gives, after a little simplification,

$$h_3 \operatorname{Im}(\bar{D}H_t) = -\partial(H/h_3)/\partial u_3 - \operatorname{Re}(H_t D^* h_3) \tag{2.20}$$

This relation, multiplied by i, can now be added to (2.15); the effect is to remove the Re operator from the left-hand side, and gives

$$h_3\bar{D}H_t = -j\omega\epsilon E - i\partial(H/h_3)/\partial u_3 - i\operatorname{Re}(H_t D^* h_3) \tag{2.21}$$

Although the Re operator has not been completely removed from the equation it has been transferred to the relatively innocuous term involving $D^* h_3$.

A similar manipulation of (2.12) and (2.15) gives

$$h_3\bar{D}E_t = j\omega\mu H - i\partial(E/h_3)/\partial u_3 - i\,\mathrm{Re}\,(E_tD^*h_3) \qquad (2.22)$$

To take the decoupling process further it is necessary to combine (2.11) and (2.12) in such a way that E_t and H_t come together in a form involving only a single combination. If (2.12) be multiplied by an undetermined multiplier N and added to (2.11) it is easy to see that E_t and H_t occur in the two combinations $E_t + NH_t$ and $\epsilon NE_t - \mu H_t$. These forms will be the same, apart from an initial factor, if N is taken as

$$N = \pm j(\mu/\epsilon)^{1/2} = \pm N_0, \quad \text{say} \qquad (2.23)$$

If we now define

$$F_{t\pm} = E_t \pm N_0 H_t \qquad (2.24)$$

$$F_\pm = E \pm N_0 H \qquad (2.25)$$

the combined equations become

$$DF = [\partial/\partial u_3 + ij\epsilon\omega h_3 N]\,F_t \qquad (2.26)$$

In this equation, which is really a pair of equations, F and F_t mean F_+ and F_{t+} when the sign of N is positive, and F_- and F_{t-} when the sign of N is reversed. The advantage of this compact form is that two relations can be treated as one, in much the same way as two real numbers x and y can be combined to form a single complex number: in the present case the separation into component parts is obtained by reversing the sign of N and adding or subtracting.

To complete the process (2.21) and (2.22) can be similarly combined, using the same multiplier N. The resulting equation can be put in the form:

$$h_3\bar{D}F_t = -[j\omega\epsilon Nh_3 + i\partial/\partial u_3](F/h_3) - i\,\mathrm{Re}\,(F_tD^*h_3) \qquad (2.27)$$

where, as with (2.26), this is really a pair of equations for the two signs of N.

2.4 Equations for the field components

To get a 'wave-equation' for the fields it is necessary to eliminate F_t between (2.26) and (2.27). The difficulty stems from the presence of the operator $\partial/\partial u_3$ since, in general, the u_3 variation will not be the usual exponential variation. The easiest place to start is (2.26) which can be integrated to give

$$F_t = e^{-i\theta} \int e^{i\theta} D(F) du_3 \tag{2.28}$$

with

$$\theta = j\epsilon\omega N \int h_3 du_3 \tag{2.29}$$

This can be inserted directly into (2.27) but it is preferable first to divide throughout by h_3 and then to operate with $[\partial/\partial u_3 + ij\epsilon\omega Nh_3]$. Since this operator does not, in general, commute with $\bar{D}$ it cannot be taken inside the $\bar{D}$ operation, where it is needed to operate directly on F_t. However, from (2.19) we get

$$\bar{D}(h_3 F_t) = h_3 \bar{D}(F_t) + iF_t D^*(h_3) \tag{2.30}$$

so, if a term $j\epsilon\omega NF_t D^* h_3$ is subtracted from each side, the result of these operations on (2.27) is

$$\bar{D}\left[\left(\frac{\partial}{\partial u_3} + ij\epsilon\omega Nh_3\right) F_t\right] = -\left(\frac{\partial}{\partial u_3} + ij\epsilon\omega Nh_3\right)$$
$$\left[\left(j\epsilon\omega N + \frac{i}{h_3}\frac{\partial}{\partial u_3}\right)\frac{F}{h_3} + \frac{i}{h_3}\,\mathrm{Re}\,(F_t D^* h_3) \;\; - j\epsilon\omega NF_t D^* \log h_3\right] \tag{2.31}$$

From (2.26) the left-hand side can now be expressed as $\bar{D}DF$ where $\bar{D}D$ is the operator of (2.16). The resulting relation can be simplified to the form

$$\left\{\nabla_t^2 + \frac{\partial}{\partial u_3}\left[\frac{1}{h_3}\frac{\partial}{\partial u_3}\frac{1}{h_3}\right] + \omega^2 \epsilon\mu\right\}F = -\,\mathrm{Re}\left[F_t\left(\frac{\partial}{\partial u_3} - 2ij\epsilon\omega Nh_3\right)\right.$$
$$\left. D^* \log h_3\right] \tag{2.32}$$

where ∇_t^2 is the transverse Laplacian. Note that the operand is $h_3(E_3 + NH_3)$. On the right F_t needs to be put in terms of F from (2.28).

The resulting equation, for F only, is accordingly an integro-differential equation, of which the dominant contribution comes from the wave equation like terms on the left. It is suitable for solution by iteration or perturbation techniques when $D^* h_3$ is small. Although rather complicated in form the equation is completely decoupled, which was an important aim of the present analysis.

To recover the field components F is found for $N = +N_0$ and $-N_0$. From (2.25), the sum of the two solutions gives $2E$ and the difference $2N_0 H$. This yields the axial components. F_t is found from (2.28) and again the sum of the two solutions gives $2E_t$ and the difference $2N_0 H_t$. Finally, from E_t we get $E_1 = \mathrm{Re}\,(E_t)$, $E_2 = \mathrm{Im}\,(E_t)$, these operations being in the field of the imaginary number i.

2.5 Uniform propagation

In special cases the propagation in the u_3 dimension may be uniform so that u_3 appears in the solution only through a factor $e^{-\gamma u_3}$, where γ is constant. Then $\partial/\partial u_3$ can be replaced by $-\gamma$ and the equations simplify greatly. Instead of (2.28) we have

$$F_t = \frac{D(F)}{-\gamma + ij\epsilon\omega h_3 N} \qquad (2.33)$$

Moreover, h_3 will be independent of u_3, so the right-hand side of (2.32) will simplify to

$$\mathrm{Re}\left\{\frac{2ij\epsilon\omega N h_3}{-\gamma + ij\epsilon\omega N h_3} D(F)D^*(\log h_3)\right\} \qquad (2.34)$$

In the present case we can also solve directly for the transverse field. If we divide (2.27) by $j\omega\epsilon N h_3 - i\gamma$ we can eliminate F to obtain

$$D\left\{\frac{h_3\bar{D}(F_t) + i\,\mathrm{Re}\,(F_t\,D^* h_3)}{-j\epsilon\omega N + i\gamma/h_3}\right\} = (-\gamma + ij\epsilon\omega h_3 N)F_t \qquad (2.35)$$

Either of these equations is suitable for iteration or solution by asymptotic or perturbation techniques when h_3 departs from unity by a small and slowly varying quantity.

It should be noted that neither $D\bar{D}$ nor $\bar{D}D$ reduces to the transverse Laplacian in the general case, though they are both second order partial differential forms.

References

1 Endean, V.G. and Allen, J.E.: 'General 3-dimensional solution of eddy-current and Laplacian fields in cylindrical structures', *IEE Contr. and Sci. Rec.*, pp. 2059–2060, Dec. 1970

2 Crout, P.D.: 'The determination of antenna patterns of n-arm antennas by means of bi-complex functions', *IEEE Trans. AP-18*, pp. 686–689, Sept. 1970

3 Lewin, L.: 'A decoupled formulation of the vector wave equation in orthogonal curvilinear coordinates, with application to ferrite-filled and curved waveguides of general cross-section', *IEEE Trans. MTT-20*, pp. 338–342, May 1972

Chapter 3

Propagation in twisted rectangular waveguides

3.1 Twisted-co-ordinate system

Fig. 3.1 shows a rectangular waveguide uniformly twisted about a straight axis. This configuration apparently cannot be handled by the techniques of Chapter 1 since the axis itself does not curve. What is needed is a co-ordinate system in which the guide cross-section is determined by natural boundaries; and a transverse rectangular grid rotating with the guide fulfills this requirement.

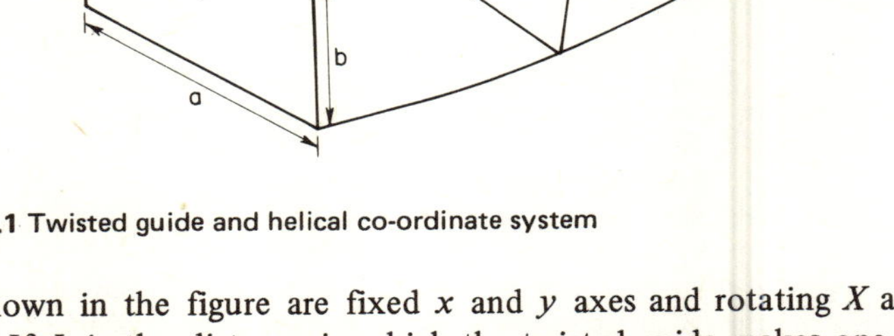

Fig. 3.1 Twisted guide and helical co-ordinate system

Shown in the figure are fixed x and y axes and rotating X and Y axes. If L is the distance in which the twisted guide makes one complete rotation, then the angular period is $p = 2\pi/L$. Accordingly the co-ordinate transformation for the twisted guide is

$$\begin{aligned} X &= x \cos pz + y \sin pz \\ Y &= y \cos pz - x \sin pz \\ Z &= z \end{aligned} \tag{3.1}$$

It is necessary to distinguish between Z and z, even though they are numerically equal, since an independent variation of z implies a basis in fixed x and y, which is not the same as fixed X and Y.

For constant Z the X,Y-axes are obtained from the x, y-axes simply by a rotation through an angle pZ. Similarly the unit vectors $\boldsymbol{a}_X$ and $\boldsymbol{a}_Y$ are related to $\boldsymbol{a}_x$ and $\boldsymbol{a}_y$ by the same rotation. However, the unit vector $\boldsymbol{a}_Z$ differs from $\boldsymbol{a}_z$ (except along the axis) owing to the fact that a Z translation, *for constant X and Y*, rotates the radius vector relative to the fixed x, y-axes. It therefore traces out a helix of radius $(x^2 + y^2)^{1/2}$ and the $\boldsymbol{a}_Z$ unit vector is tangential to this helix. Except on the z-axis this direction is not parallel to the axis, and is therefore not normal to the XY plane. The projection on the xy plane is shown in Fig. 3.2, from which it is apparent that there is a component of $\boldsymbol{a}_Z$ proportional to $x\boldsymbol{a}_y - y\boldsymbol{a}_x$. Also shown in the figure is the sinusoidal projection of the helix on the yz plane. As it crosses the axis the projection of $\boldsymbol{a}_Z$ on the transverse plane occurs at an angle θ given by the gradient of the curve $y = \sin pz$ at $z = 0$, i.e. $\tan\theta = py$. Hence

$$\boldsymbol{a}_Z = (\boldsymbol{a}_z - py\boldsymbol{a}_x + px\boldsymbol{a}_y)/M \tag{3.2}$$

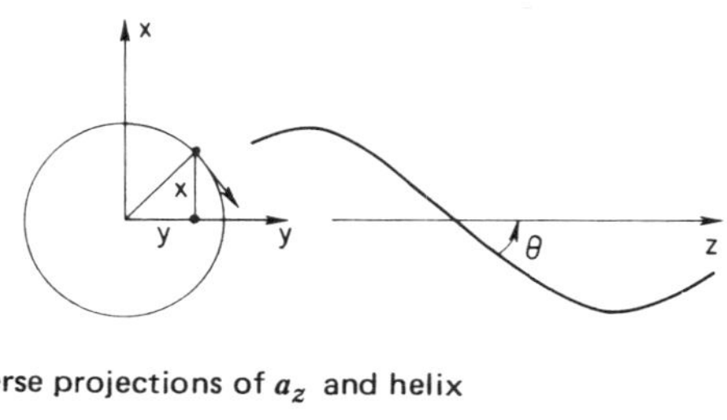

Fig. 3.2 Transverse projections of $\boldsymbol{a}_z$ and helix

where M is needed to normalize the vector to unity,

$$M^2 = 1 + p^2(x^2 + y^2) \tag{3.3}$$

It is apparent from (3.2) that $\boldsymbol{a}_Z$ is not orthogonal to $\boldsymbol{a}_x$ and $\boldsymbol{a}_y$. However, by taking the scalar product of (3.2) with $\boldsymbol{a}_z$ we get $\boldsymbol{a}_Z \cdot \boldsymbol{a}_z = 1/M$. Hence the z and Z components of any vector $\boldsymbol{A}$ are related by

$$A_Z = MA_z \tag{3.4}$$

The connection between the components of the same vector in the two co-ordinate systems is therefore

$$\boldsymbol{a}_x A_x + \boldsymbol{a}_y A_y + \boldsymbol{a}_z A_z = \boldsymbol{a}_X A_X + \boldsymbol{a}_Y A_Y + \boldsymbol{a}_Z M A_z \tag{3.5}$$

Because of the non-orthogonality of $\boldsymbol{a}_x$ and $\boldsymbol{a}_y$ to $\boldsymbol{a}_Z$ it is clear that A_x, A_y are related to A_X, A_Y by a simple rotation only in the absence of a z-component. It is also clear from (3.4) that $A_Z = 0$ when $A_z = 0$.

3.2 Field components

The electric and magnetic fields can be presented in terms of an axial vector $\boldsymbol{a}_z\psi$ through the relations[1]

$$E = \text{curl}\,(\boldsymbol{a}_z\psi)$$

$$H = [j(\epsilon/\mu)^{1/2}/k]\,[\text{grad}(\partial\psi/\partial z) + k^2\boldsymbol{a}_z\psi] \tag{3.6}$$

where $k^2 = \omega^2\epsilon\mu$ and ψ satisfies

$$\text{div grad}\,\psi + k^2\psi = 0 \tag{3.7}$$

There is also a corresponding pair of relations in which the roles of E and H are interchanged. However, (3.6) has the property $E_z = 0$ identically, and from (3.4) this implies $E_Z = 0$, i.e. zero tangential axial electric field on the guide surfaces. Since this is a boundary condition that must be satisfied for a metal guide surface, the structure of (3.6) is clearly a suitable one for the problem at hand. From (3.6), $E_x = \partial\psi/\partial y$, $E_y = -\partial\psi/\partial x$, $E_z = 0$, from which, on changing to XYZ co-ordinates

$$E_X = \partial\psi/\partial Y, \quad E_Y = -\partial\psi/\partial X, \quad E_Z = 0 \tag{3.8}$$

To express (3.7) in XYZ co-ordinates we first write it as $(\partial^2\psi/\partial x^2 + \partial^2\psi/\partial y^2) + \partial^2\psi/\partial z^2 + k^2\psi = 0$ and then transform via (3.1). The first two terms become, since only an x, y-independent rotation is involved, simply $\partial^2\psi/\partial X^2 + \partial^2\psi/\partial Y^2$. The term $\partial\psi/\partial z$ can be expressed as

$$\frac{\partial\psi}{\partial z} = \frac{\partial\psi}{\partial X}\frac{\partial X}{\partial z} + \frac{\partial\psi}{\partial Y}\frac{\partial Y}{\partial z} + \frac{\partial\psi}{\partial Z}\frac{\partial Z}{\partial z}$$

$$= pY\frac{\partial\psi}{\partial X} - pX\frac{\partial\psi}{\partial Y} + \frac{\partial\psi}{\partial Z} \tag{3.9}$$

The coefficients pY and $-pX$ come directly from differentiating (3.1).

A further differentiation with respect to z is obtained in the same way, remembering that X and Y are functions of z via (3.1). Hence the wave equation for ψ in the XYZ co-ordinates becomes

$$(1+p^2Y^2)\partial^2\psi/\partial X^2 + (1+p^2X^2)\partial^2\psi/\partial Y^2 + \partial^2\psi/\partial Z^2 + k^2\psi$$
$$+ 2p(Y\partial^2\psi/\partial X\partial Z - X\partial^2\psi/\partial Y\partial Z)$$
$$-p^2(X\partial\psi/\partial X + Y\partial\psi/\partial Y + 2XY\partial^2\psi/\partial X\partial Y) = 0 \qquad (3.10)$$

Once ψ is found from this equation, (3.8) gives the various components of the electric field, and the magnetic field can be similarly obtained from (3.6).

3.3 Dominant mode construction

We seek to construct a solution to (3.10) which reduces to the dominant mode solution $E_y = \sin(\pi x/a)$ when $p = 0$. It is not too obvious that there must be such a solution; in the case of a bent circular guide, for instance, a mode-coupling phenomenon prevents a simple reduction of the field to that of a dominant mode. However, in the present case the analysis in fact confirms the existence of such a solution. Its propagation constant in the guide depends on p, but since a change of sign of p merely alters the sense of twist, we may expect the velocity to depend explicitly on p^2 rather than p (this ignores the possibility of a dependence on $|p|$). This anticipation is also confirmed.

We hypothesize a solution of (3.10) for ψ of the form

$$\psi = e^{-j\beta Z}(\psi_0 + p\psi_1 + p^2\psi_2 + \ldots) \qquad (3.11)$$

where

$$k^2 - \beta^2 = (\pi^2/a^2)(1 + A_1 p + A_2 p^2 + \ldots) \qquad (3.12)$$

The boundary conditions to be satisfied on the guide surface are

$$E_X = 0, \quad Y = \pm b/2$$
$$E_Y = 0, \quad X = \pm a/2 \qquad (3.13)$$

together with $E_Z = 0$, the latter automatically satisfied from (3.8). In terms of ψ these conditions become

$$\partial\psi/\partial Y = 0, \quad Y = \pm b/2$$
$$\partial\psi/\partial X = 0, \quad X = \pm a/2 \qquad (3.14)$$

Substituting (3.11) into (3.10) and equating powers of p gives a set of equations for the terms ψ_n.

$$\frac{\partial^2\psi_0}{\partial X^2} + \frac{\partial^2\psi_0}{\partial Y^2} + \frac{\pi^2}{a^2}\psi_0 = 0 \qquad (3.15)$$

$$\frac{\partial^2 \psi_1}{\partial X^2} + \frac{\partial^2 \psi_1}{\partial Y^2} + \frac{\pi^2}{a^2}(\psi_1 + A_1 \psi_0) - 2jk'Y\frac{\partial \psi_0}{\partial X} = 0 \qquad (3.16)$$

$$\frac{\partial^2 \psi_2}{\partial X^2} + \frac{\partial^2 \psi_2}{\partial Y^2} + \frac{\pi^2}{a^2}\psi_2 = F(X, Y) \qquad (3.17)$$

with $k' = (k^2 - \pi^2/a^2)^{1/2}$ and

$$F(X, Y) = (A_2 - Y^2)\frac{\partial^2 \psi_0}{\partial X^2} + X\frac{\partial \psi_0}{\partial X} + 2jk'\left(Y\frac{\partial \psi_1}{\partial X} - X\frac{\partial \psi_1}{\partial Y}\right) \qquad (3.18)$$

3.4 Solution of the wave equation

Subject to the boundary conditions (3.13) the form of ψ_0 that satisfies (3.15) and gives the dominant mode in straight rectangular guide is

$$\psi_0 = \sin\frac{\pi X}{a} \qquad (3.19)$$

From (3.1) this clearly reduces, as required, to $\sin(\pi x/a)$ when $p = 0$. Substituting into (3.16) gives an equation for ψ_1

$$\frac{\partial^2 \psi_1}{\partial X^2} + \frac{\partial^2 \psi_1}{\partial Y^2} + \left(\frac{\pi^2}{a^2}\right)\left(\psi_1 + A_1 \sin\frac{\pi X}{a}\right) = 2jk'\frac{\pi Y}{a}\cos\frac{\pi X}{a} \qquad (3.20)$$

The most general form fitting the boundary conditions is

$$\psi_1 = \sum_0^\infty \sum_0^\infty A_{mn} \cos[(X - a/2)m\pi/a] \cos[(Y - b/2)n\pi/b] \qquad (3.21)$$

Insertion into (3.20) gives

$$\sum_0^\infty \sum_0^\infty \left(\frac{m^2\pi^2}{a^2} + \frac{n^2\pi^2}{b^2} - \frac{\pi^2}{a^2}\right) A_{mn} \cos\left[(X - \tfrac{1}{2}a)\frac{m\pi}{a}\right] \cos\left[(Y - \tfrac{1}{2}b)\frac{n\pi}{b}\right]$$
$$= A_1(\pi^2/a^2)\sin(\pi X/a) - 2jk'(\pi Y/a)\cos(\pi X/a) \qquad (3.22)$$

On the left of (3.22) the coefficient of $\sin(\pi X/a)$ vanishes identically, since for it, $m = 1$, $n = 0$. Hence, by the usual process for determining the coefficients of a Fourier series we find, for this term

$$0 = \int_{-a/2}^{a/2} \int_{-b/2}^{b/2} \sin(\pi X/a)[A_1 \sin(\pi X/a) - 2jk'(aY/\pi)\cos(\pi X/a)]\, dX\, dY \qquad (3.23)$$

This gives $A_1 = 0$, as anticipated, since the integration of the second term in (3.23) vanishes. With $A_1 = 0$ the coefficients A_{mn} can now be found in the usual way, and on insertion back into (3.21) give

$$\psi_1 = \sum_{m=0}^{\infty} \sum_{n=1}^{\infty} \frac{4jk'b}{\pi^2 a} \frac{\epsilon_m (1 + \cos m\pi)(1 - \cos n\pi)}{n^2 (m^2 - 1)[n^2\pi^2/b^2 + (m^2 - 1)\pi^2/a^2]} \times$$

$$\cos\left[(X - \tfrac{1}{2}a)\frac{m\pi}{a}\right] \cos\left[(Y - \tfrac{1}{2}b)\frac{n\pi}{b}\right] \tag{3.24}$$

where $\epsilon_0 = 1$, $\epsilon_m = 2$, $m > 0$. It is clear from this expression that only m even and n odd terms contribute. In particular there is no $m = 1$ term though, from (3.16), an arbitrary term in $\sin(\pi X/a)$ could apparently have been added. We shall encounter this arbitrariness in other contexts later. Its resolution can be found by examining the form of (3.11) where the amplitude of ψ_0 has been chosen as unity. Clearly any amplitude multiplying (3.11) could have been selected, and if such amplitude had been presented as a power series in p, then, on multiplying out, the effect would have been to mix in arbitrary amounts of ψ_0 into ψ_1, ψ_0 and ψ_1 into ψ_2, etc. In order to be able to work with a unique solution we can choose to *specify* that ψ_1 be linearly free of ψ_0; ψ_2 free of ψ_1 and ψ_0, etc. In this way we can then build up an unambiguous solution.

3.5 The propagation coefficient

Since A_1 in (3.11) is zero it is necessary to calculate A_2 in order to obtain the effect of the twist. As with ψ_1, ψ_2 is taken as a double Fourier series on X and Y. On substituting into (3.17) it is seen that again the coefficient of the sin $(\pi X/a)$ term vanishes. Hence, by Fourier analysis, we get from this term the relation

$$\int_{-a/2}^{a/2} \int_{-b/2}^{b/2} F(X, Y) \sin(\pi X/a)\, dX dY = 0 \tag{3.25}$$

with F given by (3.18).

On carrying out the integration we get an equation for A_2 which can be put in the form

$$(A_2 - b^2/12)\,\pi^2 b/2a - ab/4 = (32k'^2 b/\pi^4 a)S \tag{3.26}$$

where

$$S = \sum_{m=0}^{\infty} \sum_{n=1}^{\infty} \frac{\epsilon_m (1 + \cos m\pi)(1 - \cos n\pi)[b^2 m^2 (1 - m^2) + a^2 n^2 (1 + m^2)]}{n^4 (m^2 - 1)^3 [n^2\pi^2/b^2 + (m^2 - 1)\pi^2/a^2]} \tag{3.27}$$

The prime on the summation sign is to indicate that the $m = 1$ term is omitted.

In terms of A_2 the propagation coefficient β is given by

$$\beta^2 = (k^2 - \pi^2/a^2) - A_2(\pi p/a)^2 \tag{3.28}$$

A displayed form of this relation is given in (3.35).

3.6 Summation of series

The $m = 0$ term in (3.27) is

$$s_0 = -(4a^4/\pi^2) \sum_{1,3\ldots}^{\infty} [(n^2 - b^2/a^2)^{-1} - 1/n^2] \tag{3.29}$$

Now the partial fraction expansion of $\tan x$ is

$$\tan x = \sum_{1}^{\infty} \frac{2x}{(2n-1)^2\pi^2/4 - x^2} \tag{3.30}$$

Taking $x = \pi b/2a$, (3.29) sums to

$$s_0 = \tfrac{1}{2}a^4[1 - (2a/\pi b)\tan(\pi b/2a)] \tag{3.31}$$

For values of m other than 0 the general term in (3.27) is a little more complicated. Quite generally, if q is a parameter, a partial fraction expansion of $1/n^4(n^2 + q^2)$ gives

$$\frac{1}{n^4(n^2+q^2)} = \frac{1}{q^4(n^2+q^2)} + \frac{1}{q^2n^4} - \frac{1}{q^4n^2} \tag{3.32}$$

This can be used in (3.27) with $q^2 = -(m^2 - 1)b^2/a^2$ to split up the nth term into summable components. Clearly q is imaginary, so the summations involve hyperbolic tangents; the remaining terms in S become

$$-\sum_{m=2,4\ldots}^{\infty} \frac{a^4}{(m^2-1)^3}\left\{\frac{m^2\pi^2b^2}{12a^2} - \frac{2m^2+1}{m^2-1} \times \right.$$

$$\left.\left[1 - \frac{2a}{\pi b(m^2-1)^{1/2}} \tanh\left(\frac{\pi b}{2a}(m^2-1)^{1/2}\right)\right]\right\} \tag{3.33}$$

Putting $\phi = \pi b/2a$, and writing $2m$ for m, so that the new summation on m goes from 1 to ∞, it is seen that S involves an initial factor a^4 times a function of $f(\phi)$ given by

$$f(\phi) = \tfrac{1}{2}(1 - \phi^{-1}\tan\phi) + \sum_{1}^{\infty} \frac{m^2 + 1/8}{32(m^2 - 1/4)^4} \times \left\{1 - \frac{\tanh[2\phi(m^2 - 1/4)^{1/2}]}{2\phi(m^2 - 1/4)^{1/2}}\right\} \qquad (3.34)$$

Although the series in this expression is not apparently summable in closed form, a very good approximation to it can be found as follows. In the first place the summation of $(m^2 + 1/8)/(m^2 - 1/4)^4$ can be expressed in terms of the well-known Dirichlet series of inverse powers of the natural numbers, and sums to $(\pi^4/8 + 3\pi^2/4 - 16)$. For most waveguide sizes, and for m greater than 2, the hyperbolic tangents are all close to unity. A term involving one can be approximated by

$$\frac{-(m^2 + 1/8)}{64\phi(m^2 - 1/4)^4 m(1 - 1/8m^2)} \sim \frac{-m}{64\phi(m^2 - 1/4)^4(1 - 1/4m^2)}$$

$$= \frac{-m^3}{64\phi(m^2 - 1/4)^5}$$

In this form it is again summable in terms of Dirichlet series, leading to the Riemann zeta-function and giving

$$[-10 + 21\zeta(3) - 31\zeta(5)]/256\phi = -16{\cdot}902/256\phi$$

The method of summing these series is outlined in more detail in Appendix A.

Using these results an alternative form to (3.28) is

$$\frac{1}{\lambda_T^2} = \frac{1}{\lambda_g^2} - \frac{1}{L^2}\left\{\frac{1}{2} + \frac{\phi^2}{3} + \left(\frac{a^2}{\lambda^2} - \frac{1}{4}\right)\left[6 + \pi^2 - \frac{4\phi^2}{3} - \frac{g(\phi)}{\pi^2\phi}\right]\right\} \qquad (3.35)$$

where

λ_T = wavelength in twisted guide

λ_g = $\lambda/(1 - \lambda^2/4a^2)^{1/2}$, the wavelength in straight guide

ϕ = $\pi b/2a$

$$g(\phi) = 16{\cdot}902 + 128\tan\phi - 4\sum_{1}^{\infty} \frac{m^3 - (m^2 + \frac{1}{8})(m^2 - \frac{1}{4})^{1/2}\tanh[2\phi(m^2 - \frac{1}{4})^{1/2}]}{(m^2 - 1/4)^5}$$

The series is now *extremely* rapidly convergent; for most practical purposes the first term only suffices unless ϕ is unusually small. Fig. 3.3 shows the multiplier of $(a^2/\lambda^2 - 1/4)$ in (3.35) plotted against ϕ for $0 < b/a < 0{\cdot}6$.

Since the right-hand side of (3.35) is negative when $\lambda = 2a$, the effect of the twist is to increase the cutoff frequency, approximately in the ratio of $1 + (a^2 + \pi^2 b^2/6)/L^2$ to one.

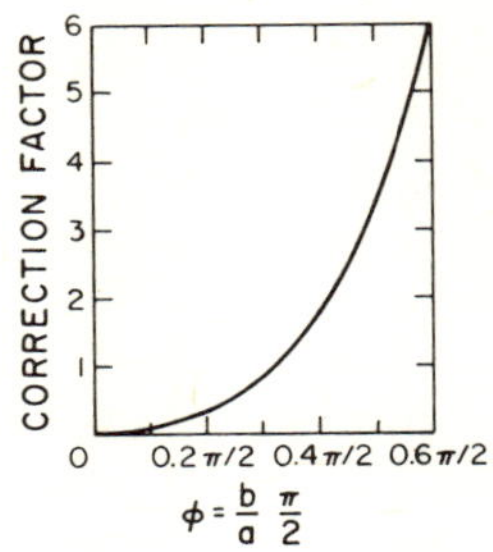

Fig. 3.3 Plot of correction factor against ϕ for twisted guide

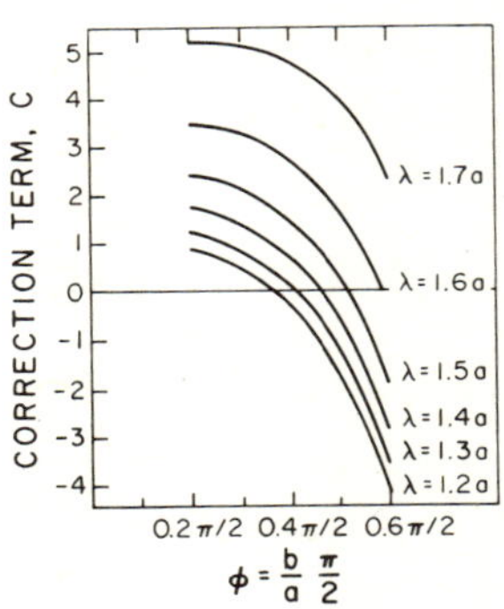

Fig. 3.4 Plot of correction coefficient againt ϕ for twisted guide

3.7 Degeneracy for square guide

The graph of Fig. 3.3 runs off to infinity as b approaches a because the tan ϕ term increases without limit for a square guide. For a narrow rectangular guide the polarization of the dominant mode twists as the guide twists. As the guide becomes more and more square in shape, there is an increasing coupling to the cross-polarized dominant mode, as may be seen from the $m = 0$, $n = 1$ term in (3.24). The square shape is degenerate in this respect, and the solution cannot therefore take the form of a single basic mode whose polarization twists with the guide. Since this assumption is really implicit in the initial choice of $\psi_0 = \sin(\pi X/a)$, this is the cause of the breakdown of (3.35) at $b = a$. It would seem that the analysis must allow for this degeneracy from the beginning by starting with two coupled modes for the initial form of solution, much as is done for the bent circular guide in Chapter 6.

Fig. 3.4 shows the correction to the straight guide wavelength for a series of values of λ/a and ϕ. The dimensionless quantity c is defined by

$$1/\lambda_T^2 = (1 - ca^2/L^2)/\lambda_g^2 \tag{3.36}$$

and is plotted against ϕ in the diagram. Although it is seen to vanish at some wavelength in the band its effect is not negligible except for twists many times longer than the guide width. And as the guide becomes more and more square the correction term increases without limit.

Reference

1 Lewin, L.: 'Theory of waveguides' (Newnes-Butterworths, London, 1975) p. 9

Chapter 4

Propagation in curved rectangular waveguides

4.1 E-plane bend: The wave equation

Fig. 4.1 shows a guide of cross-section $a \times b$ whose axis is bent to radius R, or curvature $\chi = 1/R$. The co-ordinate system is $u_1 = y$, $u_2 = x$, $u_3 = s$, with metric coefficients, from (1.19), $h_1 = h_2 = 1$, $h_3 = 1 + \chi y$. (The positive sign occurs in h_3 because y is directed in the opposite direction to the principal normal.)

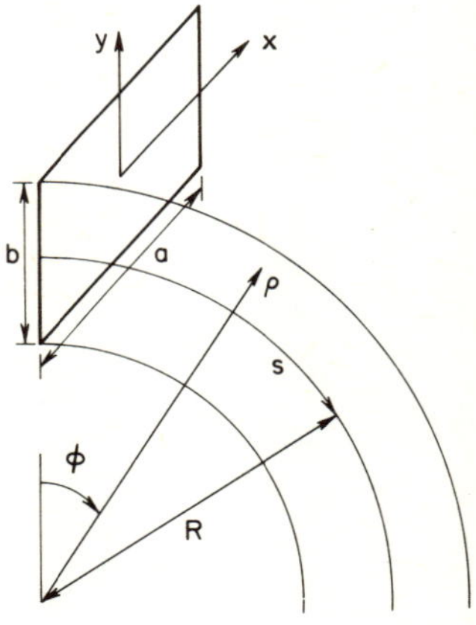

Fig. 4.1 Circular E-plane bend

The equation div grad $\psi + k^2\psi = 0$ becomes, on multiplying through by $(1 + \chi y)^2$,

$$(1 + \chi y)^2\left(\frac{\partial^2 \psi}{\partial x^2} + \frac{\partial^2 \psi}{\partial y^2}\right) + \chi(1 + \chi y)\frac{\partial \psi}{\partial y} + \frac{\partial^2 \psi}{\partial s^2} + k^2(1 + \chi y)^2\psi = 0 \tag{4.1}$$

Substantially the same equation was obtained by Jouguet,[1] in an early pioneer paper that anticipated many later developments, by transforming the wave equation in cylindrical co-ordinates, essentially by taking $\rho = R + y$ and $s = R\theta$.

In the present instance, since the guide is uniform in the x-direction, we can assume everywhere a variation as $\sin(m\pi x/a)$ for the mth mode, so that the term in $\partial^2\psi/\partial x^2$ and the last term together become $k'^2(1+\chi y)^2\psi$ where $k'^2 = k^2 - m^2\pi^2/a^2$. This is the basis for the theorem[2] that enables us to treat the guide as an ordinary parallel plate system and to replace k by k' at the close of the analysis. A certain amount of simplification is achieved thereby, and this approach will be followed here throughout the treatment of the E-plane configuration.

4.2 Perturbation solution for the dominant mode

Since[3] $\boldsymbol{H}$ can be taken equal to grad div $(\boldsymbol{a}_x\psi) + k^2\boldsymbol{a}_x\psi$, which is proportional to $\boldsymbol{a}_x\psi$ in the parallel plate case, $\boldsymbol{H}$ reduces to a single component $\boldsymbol{a}_x H_0$; hence the equations can be set up directly in terms of H_0. From $\boldsymbol{E} = -j\omega\mu$ curl $(\boldsymbol{a}_x\psi)$ it follows that the s-component of electric field, which must vanish at the guide surfaces, is proportional to $\partial H_0/\partial y$, giving

$$\partial H_0/\partial y = 0, \quad y = \pm b/2 \tag{4.2}$$

A solution of (4.1) as a power series in the curvature, assumed small, can be taken in the form

$$H_0 = Ae^{-j\beta s}[\phi_0 + \chi\xi_0 + \chi^2\zeta_0 + \ldots] \tag{4.3}$$

$$\beta^2 = k^2[1 + \chi B_1 + \chi^2 B_2 + \ldots] \tag{4.4}$$

In these expressions

A is an arbitrary amplitude;

ϕ_0 is the dominant mode form in straight guide;

ξ_0, ζ_0, etc. are functions of y, and are expansion functions pertaining to the dominant (zero-order) mode;

B_1, B_2 etc. are constants of the expansion of β in a power series in χ.

The dominant mode form ϕ_0 is a constant, and can be taken as unity, any implied amplitude being subsumed in A. As with the twisted guide of Chapter 3, a certain arbitrariness enters into the construction of

ξ_0, ζ_0, etc. since, if A contained a power series in χ and the terms were all multiplied out, an arbitrary amount of ϕ_0 could be added to ξ_0 and ϕ_0, ξ_0 added to ζ_0 etc. In order to be able to work with a unique representation we can choose to specify that ξ_0 be linearly independent of ϕ_0, ζ_0 be linearly independent of ϕ_0 and ξ_0, and so on. This is not the only choice that could be made. A requirement on normalization, for example, would yield a different representation, although it would amount merely to a redefinition of the amplitude A. Finally, since the propagation should not depend on the direction of bending, i.e. the sign of χ, we may anticipate that the constant B_1 will be zero.

Substituting (4.3) into (4.1) and comparing the coefficients of χ gives

$$d^2\xi_0/dy^2 + k^2(2y - B_1) = 0 \tag{4.5}$$

Integrating once and determining the integration constant by (4.2), which requires $d\xi_0/dy = 0$ at $y = \pm b/2$, gives

$$\xi_0' = d\xi_0/dy = k^2(b^2/4 - y^2) \tag{4.6}$$

together with, as anticipated, $B_1 = 0$.

A further integration yields

$$\xi_0 = k^2 y(b^2/4 - y^2/3) \tag{4.7}$$

No further constant of integration is needed since, as already discussed, ξ_0 is required to be linearly independent of ϕ_0, itself constant.

It can be seen from (4.7) that ξ_0 is an odd function of y.

On comparing coefficients of χ^2 when (4.3) and (4.4) are inserted into (4.1) we get

$$d^2\zeta_0/dy^2 - k^2(4y^2 + B_2 - b^2/4) + 2k^4y^2(b^2/4 - y^2/3) = 0 \tag{4.8}$$

On integrating and requiring ζ_0' to vanish at $y = \pm b/2$ we find that

$$\zeta_0' = -k^2y(b^2 - 4y^2)[1 - k^2(b^2 - y^2)/10]/3 \tag{4.9}$$

together with the condition

$$B_2 = -b^2/12 + k^2b^4/30 \tag{4.10}$$

A further integration, and excluding any constant (ϕ_0) term, gives

$$\zeta_0 = k^2y^2[2y^2 - b^2 + k^2(6b^4 - 15y^2b^2 + 8y^4)/60]/6 \tag{4.11}$$

It may be noted that ζ_0 is an even function of y.

If we define λ_E by $\beta = 2\pi/\lambda_E$ then (4.4) and (4.10) give the second order correction to the guide wavelength in the E-plane curved guide

$$\frac{1}{\lambda_E^2} = \frac{1}{\lambda^2}\,[1 - \chi^2 b^2 (1 - 2k^2 b^2/5)/12] \tag{4.12}$$

This is the parallel plate formula. For a guide of width a carrying a mode of variation $\sin(m\pi x/a)$, all of (4.7), (4.11) and (4.12) should be modified as discussed at the end of Section 4.1, with the replacement of k by k' and λ by λ_g, where $k' = 2\pi/\lambda_g$.

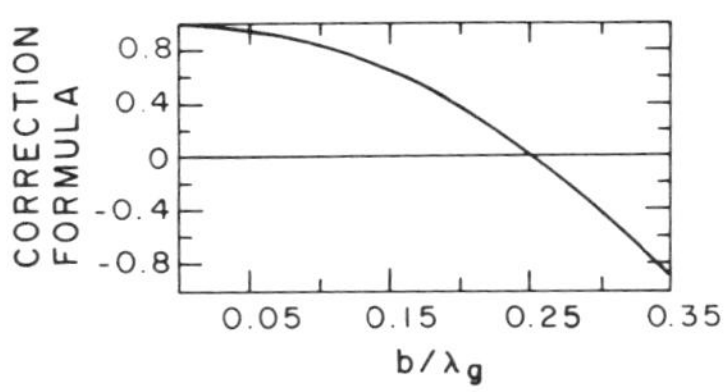

Fig. 4.2 Plot of correction function

$$f(b/\lambda_g) = 1 - \frac{2}{5}\left(\frac{2\pi b}{\lambda_g}\right)^2$$

against b/λ_g for E-plane circular bend

Although these formulae contain kb, which vanishes at long wavelengths, they are in no sense 'quasi-static' relations. In fact the kb term in (4.12) has a substantial effect. The function $1 - 2k'^2 b^2/5$ of (4.12) is shown in Fig. 4.2, plotted against b/λ_g and is seen to vanish near $b = \lambda_g/4$.

4.3 E-plane bend: higher order transverse-electric mode solution

The results of Section 4.2 for the perturbation solution are deceptively simple because of the particular form assumed by $\phi_0 = 1$. For a higher-order mode we take

$$\phi_n = \cos\left[\frac{n\pi}{b}(y - b/2)\right] \tag{4.13}$$

It satisfies

$$\phi_n'' + (k^2 - \beta_n^2)\phi_n = 0, \quad \phi_n' = 0 \quad \text{at} \quad y = \pm b/2 \tag{4.14}$$

with

$$\beta_n = (k^2 - n^2\pi^2/b^2)^{1/2} \sim -jn\pi/b \quad \text{for} \quad n \gg 1.$$

The general form of the perturbation expansion, analogous to (4.3) and (4.4), is

$$H_n = A_n e^{-j\beta(n)s}[\phi_n + \chi\xi_n + \chi^2\zeta_n + \ldots] \tag{4.15}$$

$$\beta_{(n)}^2 = \beta_n^2[1 + \chi B_{n1} + \chi^2 B_{n2} + \ldots] \tag{4.16}$$

On substituting into (4.1) and comparing the coefficients of χ we get

$$\xi_n'' + (n^2\pi^2/b^2)\xi_n + 2y\phi_n'' + \phi_n' + 2yk^2\phi_n - \beta_n^2 B_{n1}\phi_n = 0 \tag{4.17}$$

This equation can be integrated by standard techniques. The solution that satisfies $\xi_n' = 0$ at $y = \pm b/2$ and is linearly independent of ϕ_n can be verified to be

$$\xi_n(y) = (b^2/2n^2\pi^2)\{\phi_n'[\beta_n^2(y^2 - b^2/4) - k^2b^2/n^2\pi^2] - k^2y\phi_n\} \tag{4.18}$$

together with the requirement $B_{n1} = 0$.

For large n, ξ_n possess the asymptotic form

$$\xi_n \sim \tfrac{1}{2}\phi_n'(b^2/4 - y^2) \tag{4.19}$$

From (4.18), it can be seen that ξ_n is odd when ϕ_n is even, and vice versa, so that terms like $\phi_n\xi_n$ integrate to zero across the guide cross-section for all n. From the coefficients of χ^2 we get the equation

$$\zeta_n'' + (n^2\pi^2/b^2)\zeta_n + y^2\phi_n'' + 2y\xi_n'' + y\phi_n' + \xi_n' + k^2y^2\phi_n + 2k^2y\xi_n - \beta_n^2 B_{n2}\phi_n = 0 \tag{4.20}$$

This equation can be solved for ζ_n by standard techniques, though the analysis is a bit lengthy. The solution is of the form

$$\zeta_n = f_1(y)\phi_n + f_2(y)\phi_n' \tag{4.21}$$

with

$$\begin{aligned} f_1(y) &= b^2y^2[k^2(7k^2b^2/n^2\pi^2 - 4) + \beta_n^4(\tfrac{1}{2}b^2 - y^2)]/8n^2\pi^2 \\ f_2(y) &= b^4y[12k^2(7k^2b^2/n^2\pi^2 - 4) + \beta_n^2(b^2 - 4y^2) \times \\ &\quad (9k^2 - 4\beta_n^2)]/48n^4\pi^4 \end{aligned} \tag{4.22}$$

In obtaining these relations we have required $\zeta_n'(y) = 0$ at $y = \pm b/2$ and that ζ_n be linearly independent of ϕ_n. The application of the boundary conditions also leads to a unique value of B_{n2}, from which the guide wavelength is found as

$$\frac{1}{\lambda_E^2} = \frac{1}{\lambda_{gm}^2} + \frac{\chi^2}{24}\left[m^2 + \frac{12 - m^2\pi^2}{2m^2\pi^2}\left(\frac{2b}{\lambda}\right)^2 - \frac{21 + m^2\pi^2}{2m^4\pi^2}\left(\frac{2b}{\lambda}\right)^4\right], \quad m > 0 \tag{4.23}$$

with

$$\lambda_{gm} = \lambda/(1 - m^2\lambda^2/4b^2)^{1/2}$$

From (4.21) it can be seen that ζ_n is an even function of y when ϕ_n is even, and odd when ϕ_n is odd. Hence products like $\zeta_n\xi_n$ are odd, and integrate to zero over the guide cross-section. In (4.23) the transition $\lambda \to \lambda_g$ enables the result to apply to a guide of finite width.

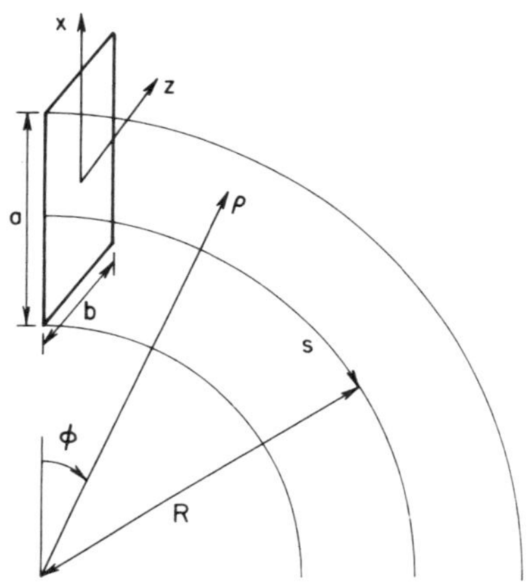

Fig. 4.3 Circular H-plane bend

4.4 H-plane bend: perturbation solution for transverse-magnetic mode

The configuration is shown in Fig. 4.3. The basic equation is the same as (4.1) except that now the radial variable is x rather than y. Instead of H_x we use E_z with boundary conditions $E_z = 0$ at $x = \pm a/2$. For the mth mode we take

$$E_m = A_m e^{-j\beta(m)s} [\phi_m + \chi\xi_m + \chi^2\zeta_m + \ldots] \quad (4.24)$$

$$\beta^2_{(m)} = \beta^2_m [1 + \chi B_{m1} + \chi^2 B_{m2} + \ldots] \quad (4.25)$$

with

$$\phi_m = \sin\left[\frac{m\pi}{a}(x - a/2)\right]$$

$$\beta_m = (k^2 - m^2\pi^2/a^2)^{1/2} \sim -jm\pi/a \quad \text{for} \quad m \gg 1. \quad (4.26)$$

The functions ϕ_m, ξ_m, ζ_m are not identical to the corresponding functions in (4.15), though they play a similar role.

The equation to be satisfied by E_m is

$$(1+\chi x)^2 \frac{d^2 E_m}{dx^2} + \chi(1+\chi x)\frac{dE_m}{dx} + (1+\chi x)^2 k^2 E_m - \beta^2_{(m)} E_m = 0 \tag{4.27}$$

If (4.24) and (4.25) are substituted into (4.27) and the coefficients of χ equated there is obtained

$$\xi''_m + (m^2\pi^2/a^2)\xi_m + 2x\phi''_m + \phi'_m + 2xk^2\phi_m - \beta^2_m B_{m1}\phi_m = 0 \tag{4.28}$$

where the primes indicate differentiation with respect to x.

As in the previous sections we require a solution that is linearly independent of ϕ_m, and satisfies, in the present instance, $\xi_m = 0$ at $x = \pm a/2$. It is readily verified that this requires $B_{m1} = 0$ and gives, similarly to (4.18),

$$\xi_m(x) = (a^2/2m^2\pi^2)[\phi'_m\beta^2_m(x^2 - a^2/4) - k^2 x\phi_m] \tag{4.29}$$

The asymptotic form for large m is

$$\xi_m(x) \sim \tfrac{1}{2}\phi'_m(a^2/4 - x^2) \tag{4.30}$$

From the coefficient of χ^2 we get the equation

$$\zeta''_m + (m^2\pi^2/a^2)\zeta_m + x^2\phi''_m + 2x\xi''_m + x\phi'_m + \xi'_m + k^2x^2\phi_m + 2k^2x\xi_m - \beta^2_m B_{m2}\phi_m = 0 \tag{4.31}$$

The solution vanishing at $x = \pm a/2$ can be found in the form

$$\zeta_m = g_1(x)\phi_m + g_2(x)\phi'_m \tag{4.32}$$

with

$$g_1(x) = -x^2[5(ka/m\pi)^4 - 2(ka/m\pi)^2 + (\tfrac{1}{2}a^2 - x^2)(a\beta^2_m/m\pi)^2]/8$$
$$g_2(x) = -x(a^2 - 4x^2)(a\beta_m/m\pi)^2(4 + 5k^2a^2/m^2\pi^2)/48 \tag{4.33}$$

together with the value

$$B_{m2} = a^4[3k^2(k^2 + 4\beta^2_m) - m^2\pi^2\beta^2_m(k^2 + 2m^2\pi^2/a^2)]/12m^4\pi^4\beta^2_m \tag{4.34}$$

From this last equation and (4.25) we get the wavelength λ_H in the H-plane curved guide in the form

$$\frac{1}{\lambda^2_H} = \frac{1}{\lambda^2_{gm}} + \frac{\chi^2}{24}\left\{m^2 - \frac{12+m^2\pi^2}{2m^2\pi^2}\left(\frac{2a}{\lambda}\right)^2 + \frac{15-m^2\pi^2}{2m^4\pi^2}\left(\frac{2a}{\lambda}\right)^4\right\} \tag{4.35}$$

with

$$\lambda_{gm} = \lambda/(1 - m^2\lambda^2/4a^2)^{1/2}$$

In particular, for the dominant mode, $m = 1$,

$$\frac{1}{\lambda_H^2} = \frac{1}{\lambda_g^2} + \frac{\chi^2}{24}\left\{1 - \frac{12+\pi^2}{2\pi^2}\left(\frac{2a}{\lambda}\right)^2 + \frac{15-\pi^2}{2\pi^2}\left(\frac{2a}{\lambda}\right)^4\right\} \quad (4.36)$$

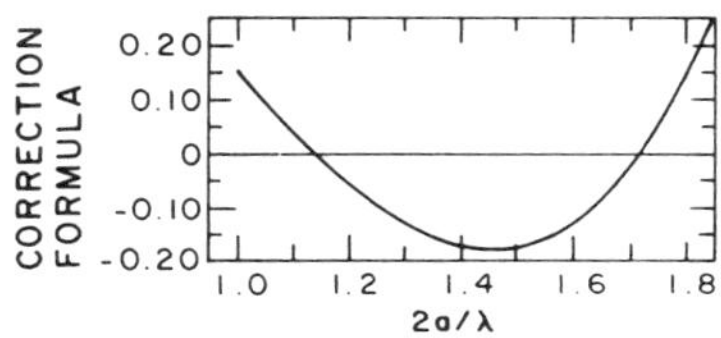

Fig. 4.4 Plot of correction function

$$g(a/\lambda) = \left[1 - \frac{12+\pi^2}{2\pi^2}\left(\frac{2a}{\lambda}\right)^2 + \frac{15-\pi^2}{2\pi^2}\left(\frac{2a}{\lambda}\right)^4\right]$$

against $2a/\lambda$ for H-plane circular bend

A plot of the coefficient of $\chi^2/24$ is given in Fig. 4.4. It reaches a minimum at $\lambda = 1{\cdot}37a$ and vanishes at $\lambda = 1{\cdot}161a$ and $1{\cdot}755a$, which lie near the extremes of the usual waveguide range.

References

1 Jouguet, M.: 'Les effets de la courbure et des discontinuités de courbure sur la propagation des ondes dans les guides á section rectangulaire', *Annales de Télécommunications,* t.2, no. 3, 1947, pp. 78–94

2 Lewin, L.: 'Theory of waveguides (Newnes-Butterworths, London, 1975) p.21

3 Lewin, L.: *loc. cit.*, p. 9

Chapter 5

Junction between straight and curved rectangular guide

5.1 Orthogonality for E-plane modes

The configuration that we shall examine is the E-plane bend of Chapter 4, and Fig. 5.1 shows the junction to a section of straight waveguide. The corresponding case for the H-bend was treated by Jouguet[1] in a pioneer paper already referred to. Owing to the extreme complexity of the analysis, he only calculated results to a first-order in the curvature. The same method used here in sections 5.8 and 5.9 can be used to deduce the second-order results for the H-bend configuration also.

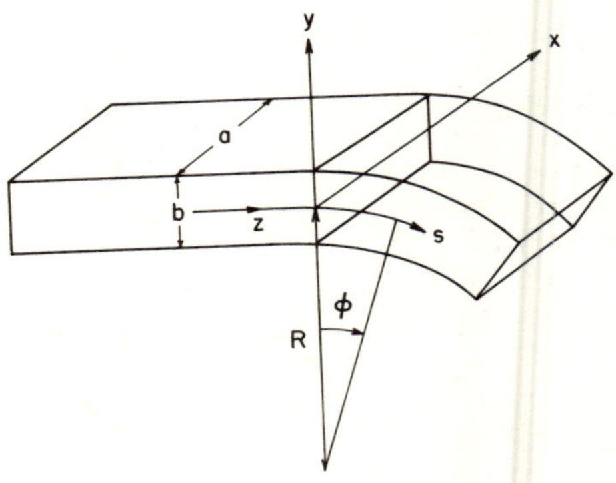

Fig. 5.1 Junction of straight and curved sections of rectangular waveguide

The fields of the nth mode can be found from the x-component of the magnetic field, the quantity H_n of (4.15). The equations satisfied by the mth and nth modes are, in the parallel plane case,

$$(1 + \chi y)^2 H_m'' + \chi(1 + \chi y)H_m' + [k^2(1 + \chi y)^2 - \beta_{(m)}^2] H_m = 0 \tag{5.1}$$

$$(1 + \chi y)^2 H_n'' + \chi(1 + \chi y)H_n' + [k^2(1 + \chi y)^2 - \beta_{(n)}^2] H_n = 0 \tag{5.2}$$

The notation is that of section 4.3. To demonstrate orthogonality, with a suitable weighting function, multiply (5.1) by H_n, (5.2) by H_m, subtract and divide by $(1 + \chi y)$. The resulting relation can be put in the form

$$\frac{\partial}{\partial y}\{(1 + \chi y)(H_m' H_n - H_n' H_m)\} = (\beta_{(m)}^2 - \beta_{(n)}^2) H_m H_n/(1 + \chi y) \tag{5.3}$$

Since H_n' and H_m' vanish at $y = \pm b/2$ the left-hand side integrates to zero over the guide cross-section, and we get

$$\int_{-b/2}^{b/2} H_m H_n (1 + \chi y)^{-1} dy = 0, \qquad m \neq n \tag{5.4}$$

It is convenient to expand H_x in the curved guide in the form

$$H_x = (1 + \chi y) \sum_0^\infty A_n \psi_n(y) e^{-j\beta_{(n)}s} \tag{5.5}$$

where $H_n = (1 + \chi y)\psi_n$.

Hence the orthogonality properties of the ψ_n are seen to be given by

$$\int_{-b/2}^{b/2} \psi_m \psi_n (1 + \chi y) dy = 0, \qquad m \neq n \tag{5.6}$$

The ψ_m are actually the eigenfunctions of the y-component of the electric field. In cylindrical co-ordinates this component amounts to the ρ-component. Since $E_\rho \propto j\rho^{-1} \partial H_x/\partial\theta$, with $\rho = R + y$ and $s = R\theta$, the expansion corresponding to (5.5) is found to be

$$E_y = (\mu/\epsilon)^{1/2} \sum_0^\infty A_n (\beta_{(n)}/k) \psi_n(y) e^{-j\beta_{(n)}s} \tag{5.7}$$

This relation confirms the interpretation of ψ_n as the mode form for the transverse component of the electric field.

5.2 Eigenfunctions in the curved guide

As the curvature goes to zero we have $H_m \to \phi_m$ where

$$\phi_m = \cos\left[\frac{m\pi}{b}(y - b/2)\right].$$

Clearly the ϕ_m are mutually orthogonal and are normalized by

$$\int_{-b/2}^{b/2} \phi_0^2 dy = b, \qquad \int_{-b/2}^{b/2} \phi_m^2 dy = b/2, \qquad m > 0 \tag{5.8}$$

With the same notation as section 4.3 we can put

$$H_n = A_n(\phi_n + \chi\xi_n + \chi^2\zeta_n + \ldots) \tag{5.9}$$

and choose the amplitude A_n to normalize H_n in the same way as ϕ_n, i.e.

$$\int_{-b/2}^{b/2} H_0^2(1+\chi y)^{-1} dy = b, \qquad \int_{-b/2}^{b/2} H_n^2(1+\chi y)^{-1} dy = b/2, \quad n > 0 \tag{5.10}$$

Since $\phi_n\xi_n$ is an odd function of y it integrates to zero over the guide, so the first terms in the curvature to contribute to (5.10) involve χ^2. Hence A_n departs from unity by a term of this order, and we can put

$$H_n = \phi_n + \chi\xi_n + \chi^2(\zeta_n + \alpha_n\phi_n) \quad \text{to order} \quad \chi^2 \tag{5.11}$$

with α_n determined by the normalization.

For the electric field eigenfunction one can similarly define

$$\psi_n = \phi_n + \chi\lambda_n + \chi^2\mu_n + \ldots \tag{5.12}$$

Then the relation $H_n = (1 + \chi y)\psi_n$ gives

$$\begin{aligned} \lambda_n &= \xi_n - y\phi_n \\ \mu_n &= \zeta_n - y(\xi_n - y\phi_n) + \alpha_n\phi_n \end{aligned} \tag{5.13}$$

These relations give the expansion terms of ψ_n in terms of ξ_n and ζ_n, which are given in (4.18) and (4.21). Note that λ_n has the same parity as ξ_n, and μ_n as ζ_n, but that neither λ_n' nor μ_n' are zero at $y = \pm b/2$.

From the orthogonality relation (5.6) one finds, on examining powers of χ and χ^2,

$$\int_{-b/2}^{b/2} (y\phi_m\phi_n + \lambda_m\phi_n + \lambda_n\phi_m) dy = 0 \tag{5.14}$$

$$\int_{-b/2}^{b/2} (\mu_m\phi_n + \mu_n\phi_m + \lambda_m\lambda_n + y\lambda_m\phi_n + y\lambda_n\phi_m) = 0 \tag{5.15}$$

In particular, with $m = 0$, $\phi_0 = 1$, (5.14) gives

$$\int_{-b/2}^{b/2} \lambda_n dy = -\int_{-b/2}^{b/2} \xi_0\phi_n dy \tag{5.16}$$

It remains to show how to calculate the normalizing constants α_n. If we insert (5.11) into (5.10) and select the coefficient of χ^2, it is found that

$$2\alpha_n \int_{-b/2}^{b/2} \phi_n^2 dy = \int_{-b/2}^{b/2} (2\phi_n \xi_n y - 2\phi_n \zeta_n - \xi_n^2 - y^2 \phi_n^2) dy \tag{5.17}$$

from which α_n can be found. In particular, with $n = 0$, $\phi_0 = 1$,

$$\alpha_0 = -\frac{b^2}{24}\left[1 - \frac{19}{30}(kb)^2 + \frac{1}{16}(kb)^4\right] \tag{5.18}$$

The method of evaluating integrals like those in (5.17) is described in Appendix B.

All these results will be found of use in succeeding sections.

5.3 Field expansions in straight guide

In the straight section $z < 0$ we take H_x for a TM wave in the form

$$H_x = \phi_0 (e^{-jkz} + \chi R_0 e^{jkz}) + 2\chi \sum_1^\infty R_n \phi_n e^{j\beta_n z} \tag{5.19}$$

This structure anticipates the conclusion that all the reflected modes will be at least $O(\chi)$ for small χ. In particular the reflected dominant mode is χR_0 so that the input impedance of the junction, normalized to that of the straight guide, is

$$Z_J = (1 - \chi R_0)/(1 + \chi R_0) \tag{5.20}$$

Similarly, the mode conversion C_n to the nth mode is

$$C_n = 2\chi R_n \tag{5.21}$$

The corresponding electric field is

$$(\epsilon/\mu)^{1/2} E_y = \phi_0 (e^{-jkz} - \chi R_0 e^{jkz}) - 2\chi \sum_1^\infty R_n (\beta_n/k) \phi_n e^{j\beta_n z} \tag{5.22}$$

At $z = 0$ the junction magnetic field H_{x0} will differ from ϕ_0 by terms at least of order χ. Let $\theta(y)$ represent this difference, so that

$$H_{x0} = \phi_0 + \theta = \phi_0 + \chi F + \chi^2 G + \ldots \tag{5.23}$$

where F and G, to be found, are the first two terms in the expansion of θ in the powers of χ. We shall here calculate properties up to and including second powers of χ only.

Using the orthogonal properties of ϕ_n, and writing the dummy integration variable as η, gives

$$R_n = (1/b) \int_{-b/2}^{b/2} (F + \chi G)\phi_n d\eta, \qquad \text{all } n \tag{5.24}$$

This can be substituted into (5.22) to give, for the junction electric field,

$$(\epsilon/\mu)^{1/2} E_{y0} = \phi_0 \left[1 - (\chi/b) \int (F + \chi G) d\eta\right] - (2\chi/kb) \sum_1^{\infty} \beta_n \phi_n \int (F + \chi G)\phi_n d\eta \tag{5.25}$$

From here to the end of this chapter all integrations will be over the guide cross-section from $-b/2$ to $b/2$, and this range will be implicit in such expressions as (5.25).

5.4 Field expansions in curved guide

The forms to be used are those of (5.5) and (5.7). For $s > 0$ we have

$$H_x = (1 + \chi y)\left\{T_0 \psi_0 e^{-j\beta s} + 2 \sum_1^{\infty} T_n \psi_n e^{-j\beta_{(n)} s}\right\} \tag{5.26}$$

From (4.4), by taking the square root,

$$\beta = k(1 + \tfrac{1}{2}B_2\chi^2) \quad \text{to} \quad 0(\chi^2) \tag{5.27}$$

with B_2 given by (4.10).

From (4.16) we similarly have

$$\beta_{(n)} = \beta_n(1 + \tfrac{1}{2}B_{n2}\chi^2) \quad \text{to} \quad 0(\chi^2) \tag{5.28}$$

However, T_n turns out to be $0(\chi)$, so the correction to β_n in (5.29) would give terms of $0(\chi^3)$. These will not be retained in the present analysis so that, in general, the difference between $\beta_{(n)}$ and β_n can be neglected for $n > 0$. The one exception to this will be considered later in the discussion of properties close to a higher-order mode cut-off.

The electric field is given by (5.7),

$$(\epsilon/\mu)^{1/2} E_y = (\beta/k) T_0 \psi_0 e^{-j\beta s} + (2/k) \sum_1^{\infty} T_n \psi_n \beta_{(n)} e^{-j\beta_{(n)} s} \tag{5.29}$$

At $s = z = 0$, Fourier analysis of the magnetic field over the set ψ_n gives, to $0(\chi^2)$,

$$T_n = \frac{1}{b}\int(\phi_0 + \chi F + \chi^2 G)\psi_n d\eta \tag{5.30}$$

As already indicated, these integrations are from $-b/2$ to $+b/2$. Because of the orthogonality of ϕ_0 with ϕ_n, the leading term of ψ_n, (5.30) will be at least $0(\chi)$.

Substituting (5.30) into (5.29) and taking $s = 0$ gives the junction electric field in the curved guide,

$$(\epsilon/\mu)^{1/2}E_{y0} = (\beta/kb)\psi_0\int(\phi_0 + \chi F + \chi^2 G)\psi_0 d\eta +$$
$$(2/kb)\sum_1^\infty \beta_{(n)}\psi_n\int(\phi_0 + \chi F + \chi^2 G)\psi_n d\eta \tag{5.31}$$

5.5 Integral equation for the junction field

To order χ^2 we can now get an integral equation from which F and G can be calculated by equating (5.25) to (5.31), since they both represent the same junction field. The range of both the field variable y and the integration variable η is from $-b/2$ to $+b/2$. Hence, to $0(\chi^2)$,

$$k\left\{b - \chi\int(F + \chi G)d\eta - \psi_0(1 + \tfrac{1}{2}B_2\chi^2)\int(1 + \chi F + \chi^2 G)\psi_0 d\eta\right\} =$$
$$2\sum_1^\infty \beta_n\left\{\phi_n\int(F + \chi G)\phi_n d\eta + \psi_n\int(1 + \chi F + \chi^2 G)\psi_n d\eta\right\} \tag{5.32}$$

In this equation we have taken $\beta_{(n)} = \beta_n$ for $n > 0$, and also put $\phi_0 = 1$. Substituting for ψ_0 from (5.12), (note that here λ_0 and μ_0 are expansion functions and do not represent wavelength or permeability), and again retaining only terms to $0(\chi^2)$, the left-hand side of (5.32) reduces to

$$-kb\chi\lambda_0 - k\chi^2\left[b(\mu_0 + \tfrac{1}{2}B_2) + \int(\mu_0 + F\lambda_0 + 2G)d\eta\right].$$

In performing this calculation we have used the fact that λ_0 and also F (see later) are odd functions of η and integrate to zero over the cross-section.

The right-hand side of (5.32) can be similarly reduced by substituting for ψ_n via (5.12), keeping terms up to $0(\chi^2)$, and also using (5.16) and (5.15) with $m = 0$ to eliminate terms in $\int\lambda_n d\eta$ and $\int\mu_n d\eta$.

Finally, by equating the coefficients of χ and χ^2 there is obtained the pair of equations

$$kb\lambda_0(y) + 2\sum_1^\infty \beta_n\phi_n(y)\int[2F(\eta) - \xi_0(\eta)]\,\phi_n(\eta)d\eta = 0 \qquad (5.33)$$

$$kb\,[\mu_0(y) + \tfrac{1}{2}B_2] + k\int[2G(\eta) + F(\eta)\lambda_0(\mu) + \mu_0(\eta)]\,d\eta =$$

$$2\sum_1^\infty \beta_n\left\{\phi_n(y)\int\left\{\phi_n(\eta)[\zeta_0(\eta) - 2G(\eta)] + \lambda_n(\eta)[\xi_0(\eta) - F(\eta)]\right\}d\eta +\right.$$

$$\left.\lambda_n(y)\int\phi_n(\eta)\,[\xi_0(\eta) - F(\eta)]\,d\eta\right\} \qquad (5.34)$$

Both equations hold for $-b/2 < y,\ \eta < b/2$.

5.6 First-order solution

If we put $2F(\eta) - \xi_0(\eta) = f(\eta)$, (5.33) becomes

$$kb\lambda_0(y) + 2\sum_1^\infty \beta_n\phi_n(y)\int_{-b/2}^{b/2} f(\eta)\phi_n(\eta)d\eta = 0, \quad -b/2 < y < b/2 \qquad (5.35)$$

If we made the approximation $\beta_n \sim -jn\pi/b$, this equation could be presented as a singular integral equation and solved, to a quasi-static approximation, in closed form. However, in the present instance we can do better than this since the $\phi_n(y)$ in (5.35) are orthogonal over the range $-b/2 < y < b/2$. This is not the usual case in waveguide problems; for instance the presence of a diaphragm at the junction would destroy this equality of the orthogonality range to the range of validity of (5.35).

If we assume a Fourier expansion in the form $f(\eta) = \sum_0^\infty A_m\phi_m(\eta)$ and insert into (5.35), a straightforward Fourier analysis gives

$$A_m = -(2k/b\beta_m)\int_{-b/2}^{b/2} \lambda_0(y)\phi_m(y)dy \qquad (5.36)$$

The integration is discussed in Appendix B. We define two functions of m of frequent occurrence here by

$$M = (1 - \cos m\pi)\,(1 - 2k^2b^2/m^2\pi^2)/m^3, \qquad m > 0 \quad (5.37)$$

$$N = m\pi/jb\beta_m = (1 - k^2b^2/m^2\pi^2)^{-1/2}, \qquad m > 0 \quad (5.38)$$

Note that $M = 0$ for m even, and $N \sim 1 + k^2b^2/2m^2\pi^2$ for $m \gg 1$. The quasi-static form comes by taking $N = 1$ for $m > 0$.

In terms of these quantities (5.36) gives

$$A_m = (2jkb/\pi^3)MN, \qquad m > 0 \qquad (5.39)$$

Moreover, A_0 from (5.36) is zero since $\lambda_0(y)$ is an odd function of y. Hence all the A_m are zero when m is even, and it is readily found that

$$F(\eta) = \tfrac{1}{2}\xi_0(\eta) + (jkb^2/\pi^3)\sum_1^\infty MN\phi_m(\eta) \qquad (5.40)$$

Since $\xi_0(\eta)$ and $\phi_m(\eta)$ for m odd are both odd functions of η, it follows that $F(\eta)$ is also an odd function, a result that was used in anticipation in the previous section.

Although not in closed form, (5.40) is exact. As already discussed, the quasi-static approximation comes from putting $N = 1$, when the equation can be summed in terms of Clausen functions,[2] but there is no need to do this here.

From (5.24) we get the first-order expression for the mode conversion.

$$C_m = 2\chi R_m = (\chi kb^2/\pi^3)\,[jMN + 2(1-\cos m\pi)kb/\pi m^4] \qquad (5.41)$$

This is clearly zero when m is even, so the even-order conversion coefficients, including, particularly, the reflection coefficent, are at least $0(\chi^2)$. For m large (5.41) gives

$$C_m \sim j2\chi kb^2/m^3\pi^3, \qquad m \text{ odd and} \gg 1 \qquad (5.42)$$

so that the coupling into the odd-order modes varies as the inverse cube of the mode number for large order.

5.7 Behavior near cut-off

At the cut-off of the mth mode $\beta_m \to 0$ and the corresponding value of N becomes infinite. Hence (5.40) for the first-order junction field, together with any formulae derived from it, such as (5.41) for the mode conversion, also become infinite at cut-off. This is clearly an impermissible behavior, and stems, in the first instance, from neglecting the difference between β_m and $\beta_{(m)}$. The term giving rise to this undesirable effect really involves, not $1/\beta_m$, but $2/(\beta_m + \beta_{(m)})$. From (4.16) and (4.23) the relation between $\beta_{(m)}$ and β_m can be put in the form

$$\beta_{(m)}^2 = \beta_m^2 + \chi^2 K_m, \quad K_m = \beta_m^2 B_{m2} \qquad (5.43)$$

where K_m remains finite and non-zero at cut-off. The trouble comes in taking the square root of (5.43) in the form $\beta_{(m)} \approx \beta_m + \chi^2 K_m/2\beta_m$. These are the first two terms of a binomial expansion of radius of convergence $\chi^2 K_m/\beta_m^2 = 1$, and although convergent for χ small enough, the range of χ for convergence shrinks to zero at cut-off.

However, we can also write $2/(\beta_m + \beta_{(m)})$ in the form $2(\beta_{(m)} - \beta_m)/(\beta_{(m)}^2 - \beta_m^2) = 2(\beta_{(m)} - \beta_m)/\chi^2 K_m$. This is clearly finite, even at cut-off, where it reaches its maximum value of $1/|\chi K_m^{1/2}|$. Hence the first-order field $F(\eta)$ and the odd-order mode conversion, both of which contain a factor χ, become $0(1)$ rather than infinite close to cut-off. Note that the curvature term entering these formulae is $|\chi|$ rather than χ, thus preserving the character of the results when the direction of curvature is changed.

The replacement of $1/\beta_m$ by $2(\beta_{(m)} - \beta_m)/\chi^2 K_m$ in the relevant formulae is to be taken as implicit when results in the neighborhood of cut-off are concerned.

5.8 The reflection coefficient

The main mode reflection χR_0 is given via (5.24) with $n = 0$, and since F is an odd function of η and integrates to zero we get

$$\chi R_0 = (\chi^2/b)\int_{-b/2}^{b/2} G(\eta)d\eta \tag{5.44}$$

Accordingly we are led to a consideration of solving (5.34) for G. This can in fact be done, but the analysis is extremely heavy, tedious and lengthy. As it happens, by a judicious use of (5.32) the even mode conversion coefficients, including R_0, can be found by using only the properties of $F(\eta)$. From (5.20) for the junction impedance it is possible to use a variation procedure using $1 + \chi F(\eta)$ as a trial function. The departure from the exact form is then of order $\chi^2 G(\eta)$, to $0(\chi^2)$, and because of the variational form the error introduced will be $0(\chi^4)$. Thus the method will provide *exact* information on the reflection coefficient to $0(\chi^2)$, which is all that could have been extracted from $G(\eta)$ anyway. Thus the determination of G from (5.34) is unnecessary.

In (5.25) and (5.31) replace $1 + \chi F + \chi^2 G$ by $1 + \theta(\eta)$ from (5.23), so that these equations are now exact. Equate these two expressions for the junction electric field, multiply by $1 + \theta(y)$, integrate across the guide cross-section and divide through by $b^2(1 + \chi R_0)^2 = [\int(1 + \theta)dy]^2$. The resulting relation, which is both exact and in variational form, is

$$Z_J = \frac{\beta[\int(1+\theta)\psi_0 dy]^2 + 2\sum_1^\infty\{\beta_n[\int(1+\theta)\phi_n dy]^2 + \beta_{(n)}[\int(1+\theta)\psi_n dy]^2\}}{k[\int(1+\theta)dy]^2} \tag{5.45}$$

Now take $\theta = \chi F$ so that, as already indicated, (5.45) will be correct to $0(\chi^2)$. Since $\psi_n = \phi_n + 0(\chi)$ and $\int \phi_n dy = 0$ for $n > 0$, all the squared integrals in the series will be $0(\chi^2)$. Hence the difference between $\beta_{(n)}$ and β_n (except in the neighborhood of cut-off, when the remarks of section 5.7 apply) can be ignored. Using (5.12) for ψ_n and keeping terms up to χ^2 gives

$$Z_J = (\beta/kb)\left[b + 2\chi^2 \int (F\lambda_0 + \mu_0)dy\right] + (2\chi^2/kb^2)\sum_1^\infty \beta_n \left\{\left[\int F\phi_n dy\right]^2 + \left[\int (F\phi_n + \lambda_n)dy\right]^2\right\} \tag{5.46}$$

It remains to substitute for μ_0 via (5.13), ξ_0 and ζ_0 via (4.7) and (4.11), F via (5.40), λ_n via (5.16) and carry out the indicated integrations. The method of dealing with the latter is discussed in Appendix B. The final result of these operations is

$$Z_J = \frac{\beta}{k}\left[1 + \frac{b^2\chi^2}{12}\left(1 - \frac{k^2b^2}{5}\right)\right] - j\frac{2k\chi^2b^3}{\pi^5} \times \sum_{\substack{1 \\ (m\ \text{odd})}}^{\infty} \frac{1}{m^5(1 - k^2b^2/m^2\pi^2)^{1/2}} \tag{5.47}$$

The first term can be further simplified in view of (4.12) connecting $\beta^2 = (2\pi/\lambda_E)^2$ with k^2 and B_2. It represents the (normalized) junction resistance R_J and to $0(\chi^2)$ can be put in the form

$$R_J = 1 + \chi^2 b^2/24 \tag{5.48}$$

It will be noticed that this is somewhat different from the simple ratio of the guide wavelengths which, from (4.12) is equal to $1 + \chi^2 b^2(1 - 2k^2b^2/5)/24$. The junction therefore exhibits a transformer effect; the turns ratio n, to $0(\chi^2)$ is readily found, and is given by

$$n = 1 + \chi^2 k^2 b^4/120 \tag{5.49}$$

An equivalent circuit for the junction is given in Fig. 5.2.*

The reason for the presence of this transformer effect is that the form of the dominant mode in the curved guide is not exactly constant (as is $\phi_0 = 1$), so that the coupling into it from the straight guide differs somewhat from unity.

*The parallel capacity gives the effect of the negative capacitive term of (5.47) for a nearly matched termination. It can be tuned out over a broad band by an inductive insert.

It is, perhaps, worth emphasizing that all of (5.41), (5.47), (5.48) and (5.49) are exact to $0(\chi^2)$. Close to cut-off of a mode the remarks of section 5.7 apply to the coefficient associated with that mode.

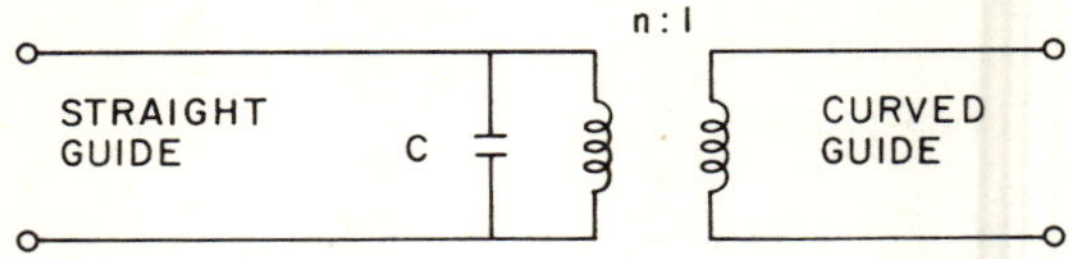

Fig. 5.2 Junction equivalent circuit

5.9 Even-order modes

A similar device to that used in section 5.8 can be used to find the even order mode-conversion coefficient $2\chi R_m$, although the expressions obtained are not necessarily in variational form. Multiply (5.25) and (5.31) by $(1+\chi F+\chi^2 G)\phi_m$, where m is even, integrate over the cross-section, and equate. Since $2\chi R_m = (2\chi^2/b)\int G\phi_m dy$ an expression for $(2\chi R_m)$ appears involving, among other terms, integrals in G. It turns out, however, that to order χ^2 the only ones of these that survive are themselves multiples of χR_m so that the entire expression gives the mode conversion to order χ^2 in terms of integrals of F and known quantities. Hence the evaluation can be made without a need to know the explicit form of G. The resulting equation, to order χ^2, is

$$C_m = 2\chi R_m = -(\chi^2/b^2\beta_m)\Bigg\{\beta_m b\int(F\lambda_m+\mu_m)d\eta + kb\times \int(\lambda_0 F+\mu_0)\phi_m\,d\eta + \sum_{\substack{1\\ n\text{ odd}}}^{\infty} 2\beta_n\Bigg[\int F\phi_n d\eta\int F\phi_n\phi_m d\eta + \int(F\phi_n+\lambda_n)d\eta\times \int(F\phi_n+\lambda_n)\phi_m d\eta\Bigg]\Bigg\} \tag{5.50}$$

The integrations can be performed on substituting for F from (5.40), and making extensive use of (5.13), (4.17), (4.18), (5.14), and (5.15). A considerable amount of manipulation and summation of series is needed to get the result in the final form

$$C_m = -(\chi^2/\beta_m)\Bigg\{\beta_m\Bigg[\frac{2b^2\phi^2}{\pi^2 m^4}\left(1-\frac{7\phi^2}{4m^2}-\frac{\pi^2\phi^2}{12}\right) + \frac{4\phi b}{\pi^3}\sum_{\substack{1\\ n\text{ odd}}}^{\infty}\frac{(1-2\phi^2/n^2)[2\phi^2(m^2+n^2)-n^2(n^2+3m^2)]}{n^2\beta_n(n^2-m^2)^3}\Bigg] +$$

$$+\frac{\phi^2}{2km^2}\left[1+\frac{\phi^2}{2m^2}-\frac{\pi^2\phi^4}{3m^2}-\frac{73\phi^4}{m^4}\right]+$$

$$\left.\frac{16\phi^2}{\pi^2}\sum_{\substack{1\\(n\text{ odd})}}^{\infty}\frac{n^2m^2(m^2+3n^2)-\phi^2(2n^4+5n^2m^2+m^4)+2\phi^4(m^2+n^2)}{n^4\beta_n(n^2-m^2)^3}\right\}$$

$$(5.51)$$

where $\phi = kb/\pi, m$ is even, and the summations are taken over n odd. The summations are discussed in Appendix A and the integrations in Appendix B.

The rather formidable expression in (5.51) for $C_m = 2\chi R_m$ makes it apparent why the solution for G, which is a Fourier summation over R_m, is so cumbersome to obtain directly.

For large m it is the $1/m^2$ terms in the brackets which are dominant, and give

$$C_m \sim -(\chi^2 b^2\phi/2\pi^2 m^3)\,\{j+(\phi/\pi^2)[28\zeta(3)-31\zeta(5)\phi^2]\} \quad (5.52)$$

Here, $\zeta(r)$ is the Riemann zeta-function. Clearly $C_m \sim m^{-3}$, as was also the case for the odd modes.

5.10 The junction field

The magnetic field at the junction is a Fourier summation over R_m, with the coefficients taken from (5.41) and (5.51). The summation cannot be achieved in terms of conventional functions, though the use of the form of the coefficients for large m gives information on the structure of the field near the boundaries. The third-order Clausen function is $Cl_3(x) = \sum_{q=1}^{\infty} \cos(qx)/q^3$, and has been studied extensively[2]. Its second derivative $d^2Cl_3(x)/dx^2 = \log|2\sin\frac{1}{2}x|$ and possesses a logarithmic singularity at $x = 0, 2\pi$. In terms of this function the functional form of the junction magnetic field near the guide wall contains a constitutent, obtained from the form of the terms for large m, of

$$H \sim \frac{j\chi b\phi}{\pi^2}\left\{Cl_3\left(\frac{\pi}{2}-\frac{\pi y}{b}\right)-\frac{1}{8}Cl_3\left(\pi-\frac{2\pi y}{b}\right)\right\}$$

$$-\frac{\chi^2b^2\phi}{16\pi^2}\left\{j+\frac{\phi}{\pi^2}[28\zeta(3)-31\zeta(5)\phi^2]\right\}Cl_3\left(\pi-\frac{2\pi y}{b}\right) \quad (5.53)$$

This may be used to examine the form of the field near $y = \pm b/2$. Clearly it is finite there, but its second derivative possesses a logarithmic singularity. The amount of singularity differs at the top and bottom wall for the first-order curvature term, but is the same for the second-order term.

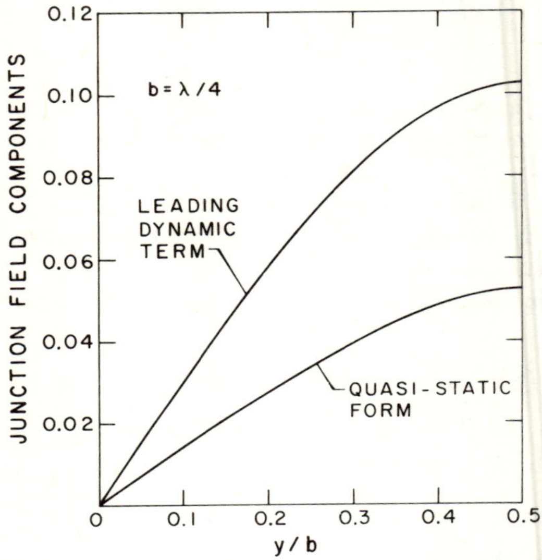

Fig. 5.3 First-order curvature perturbation to junction magnetic field

Apart from the dominating constant field component (normalized here to unity), that term in (5.53) linear in the curvature is, in fact, the quasi-static approximation to the junction field, so that the field variation, to this order in kb, can be readily extracted from (5.53). Apart from the dimensionless factor $j\chi b$, it is shown in Fig. 5.3 for the case $b = \lambda/4$. From (5.40) or (5.41) the field also contains a higher-order component $\frac{1}{2}\chi\xi_0(y) = \frac{1}{2}\chi k^2 y(b^2/4 - y^2/3)$. Although negligible in the limit $kb \to 0$ it is far from negligible when $b = \lambda/4$, and apart from a factor χb is also shown in the figure.

5.11 Concluding remarks

Almost nothing is known about the convergence of the basic series forms (5.23), (4.3) or (4.4). They appear as if they might be asymptotic expansions in $1/R$ or χ. The comments in section (5.7) show that near cut-off the convergence is necessarily limited by the ordinary binomial expansion of $\beta_{(m)}$, though this was avoided by an artifice which essentially allowed $F(\eta)$ to contain a term in χ. Now the form (5.23) was chosen so that $F(\eta)$ was the coefficient of χ in an expansion and rather implied that it was independent of χ. However, there is nothing in the analysis which obliges this to be so provided appropriate higher-order terms in the χ expansion are retained. The analysis given here is self-consistent up to terms in χ^2, but the anomalous behavior of individual terms close to cut-off, it could be argued, calls for extra terms to then be considered. It would seem, however, that when a term is near cut-off and is abnormally large, it completely dominates the scene, and makes it unnecessary to consider a higher-order expansion term which ordinarily would be completely negligible anyway.

The methods of this chapter apply to both the E- and H-bends, and can be used for any incident exciting mode. This was, in fact, done by Jouguet[1] in the paper referred to earlier, but because of the complexity of the analysis, only terms up to order χ were considered. The methods of sections 5.8 and 5.9 show how the χ^2 terms can also be extracted from the first-order solutions, and this aspect of the method should have a fairly general field of application. Moreover, in a 1950 paper, Rinow[3] showed that these techniques can be expanded to guides of arbitrary cross-section, and presented specific results for a circular coaxial waveguide.

References

1 Jouguet, M.: 'Les effets de la courbure et des discontinuités de courbure sur la propagation des ondes dans les guides á section rectangulaire', *Annales des Télécommunications*, t.2, No. 3, 1947, pp. 78–94

2 Lewin, L.: 'Dilogarithms and associated functions' (MacDonald, London, 1958) Chapter 7

3 Rinow, W.: 'Uber eine Anwendung der Störungsrechnung das Problem der gekrümmten Leitung', *Math. Nachrichten*, v.3, 1950, pp. 176–192

Chapter 6

Propagation in curved circular waveguide

6.1 Effect of mode degeneracy on perturbation solutions

Curved circular waveguide was investigated by Jouguet[1] in a pioneer paper published in 1947. He was interested in the TE_{01} mode and showed that it did not, as might have been expected of a solution perturbed in a minor way by the curvature, propagate more or less unchanged round the bend. Rather, it broke up into the sum of two distinct modes, resembling the sum and difference of the TE_{01} mode with the TM_{11} mode. Since these combinations propagated with different velocities in the curved section, their recombination at any cross-section involved a phase change depending on the angle of bend. At a certain critical angle the phase change can be 180°, and the recombination produces not the TE_{01} mode but the TM_{11} structure. This behavior is quite different from that of the TE_{01} mode in rectangular guide, treated in Chapter 4, where the mode structure is substantially unchanged as it propagates around the bend.

The reason is to be found in the degeneracy of the TE_{01} mode with the TM_{11} mode in straight circular guide. Since they have the identical velocity, *any* linear combination of the two modes will itself behave as a single mode. It is not obvious what, if any, linear combination has more significance than any other. It is perhaps 'natural' to think of the TE and the TM series as basic, but it is really the mode *stability* that should be used to determine the matter. For example, Karbowiak[2] has shown that in rectangular guide, where the TE and TM mode series for the same set of mode numbers are degenerate, it is the LS (longitudinal section) modes that are stable to perturbations caused by slight wall losses. In the circular waveguide case a certain combination of TE_{01} and TM_{11} (the latter with the direction of electrical polarization aligned perpendicular to the plane of the bend) is stable to bending, and becomes the natural combination for consideration. Actually, there are two such combinations, with opposite signs of TM_{11}, and these combinations travel with different velocities in the curved guide, hence giving rise to the phase difference already referred to.

Accordingly, the perturbation calculations need to be made on these modal combinations. At the entrance or exit of the bend they then themselves combine to produce a mix of TE_{01} and TM_{11}, as discovered by Jouguet.

6.2 The wave equation in curved co-ordinates

The transverse (local) co-ordinates are ρ and ϕ, and with the guide curved as in Fig. 6.1 the metric coefficients become, from (1.25),

$$h_\rho = 1, \quad h_\phi = \rho, \quad h_s = 1 - \chi\rho\cos\phi = h, \quad \text{say.} \tag{6.1}$$

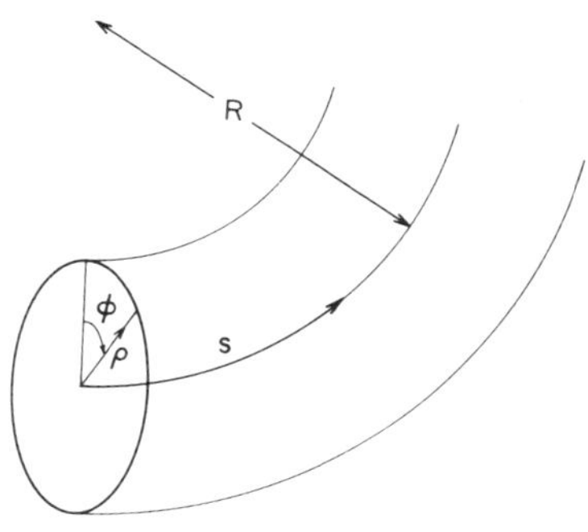

Fig. 6.1 Curved circular waveguide

Here, $\chi = 1/R$ is the curvature, and will be assumed small in the ensuing sections. Following the method detailed in Chapter 2, Maxwell's equations become, on putting $H = hH_s, E = hE_s$

$$\begin{aligned} \frac{1}{\rho h}\left[\frac{\partial E}{\partial \phi} - \rho\frac{\partial E_\phi}{\partial s}\right] &= -jk\hat{\xi}H_\rho \\ \frac{1}{h}\left[\frac{\partial E_\rho}{\partial s} - \frac{\partial E}{\partial \rho}\right] &= -jk\hat{\xi}H_\phi \end{aligned} \tag{6.2}$$

$$\frac{h}{\rho}\left[\frac{\partial(\rho E_\phi)}{\partial \rho} - \frac{\partial E_\rho}{\partial \phi}\right] = -jk\hat{\xi}H \tag{6.3}$$

$$\frac{1}{\rho h}\left[\frac{\partial H}{\partial \phi} - \rho\frac{\partial H_\phi}{\partial s}\right] = jk\hat{\eta}E_\rho$$

$$\frac{1}{h}\left[\frac{\partial H_\rho}{\partial s} - \frac{\partial H}{\partial \rho}\right] = jk\hat{\eta}E_\phi \qquad (6.4)$$

$$\frac{h}{\rho}\left[\frac{\partial(\rho H_\phi)}{\partial \rho} - \frac{\partial H_\rho}{\partial \phi}\right] = jk\hat{\eta}E \qquad (6.5)$$

$\hat{\xi} = (\mu/\epsilon)^{1/2}$, and $\hat{\eta} = (\epsilon/\mu)^{1/2}$ are the wave impedance and admittance of the material in the guide. Because h_s does not contain s we can seek a uniformly propagating solution in which all terms vary as $e^{-\gamma s}$, so that $\partial/\partial s$ can be replaced by $-\gamma$. This enables the analysis to be simplified considerably. Define, as in (2.9), transverse combinations of components, in the field of the complex variable i, by

$$E_t = E_\rho + iE_\phi, \quad H_t = H_\rho + iH_\phi \qquad (6.6)$$

Define also the operators D, D^* as in (2.10). Then (6.2) and (6.4) can be put in the form

$$D(E) + \gamma E_t = -ijkh\hat{\xi}H_t$$
$$D(H) + \gamma H_t = ijkh\hat{\eta}E_t \qquad (6.7)$$

Eliminating in turn H_t and E_t between these equations gives

$$\Gamma^2 E_t = -\gamma D(E) + ijkh\hat{\xi}D(H)$$
$$\Gamma^2 H_t = -\gamma D(H) - ijkh\hat{\eta}D(E) \qquad (6.8)$$

with

$$\Gamma^2 = k^2h^2 + \gamma^2 \qquad (6.9)$$

Note that Γ contains the transverse coordinates via h, from (6.1), and is therefore not the transverse propagation 'constant', though it is related to it.

To proceed with the analysis we can use a property of the D^* operator in circular cylindrical coordinates. It is readily verified that

$$-\text{Re}\, iD^*(\rho E_t) = \partial(\rho E_\phi)/\partial\rho - \partial E_\rho/\partial\rho \qquad (6.10)$$

with an identical relation for H_t. The right-hand side is the form occurring in (6.3) and (6.5). Use of this result enables us to avoid the use of the $\bar{D}$ operator of section 2.3. Substituting E_t and H_t from (6.8) into (6.10), and referring to (6.3) and (6.5) gives

$$-jk\rho\hat{\xi}H = h\,\text{Re}\,D^*\{\rho\Gamma^{-2}[i\gamma D(E) + jkh\hat{\xi}D(H)]\} \qquad (6.11)$$

$$jk\rho\hat{\eta}E = h\,\text{Re}\,D^*\{\rho\Gamma^{-2}[i\gamma D(H) - jkh\hat{\eta}D(E)]\} \qquad (6.12)$$

These relations constitute a pair of *coupled* equations for E and H.

It is readily verified that

$$D^*[\rho D(f)] = \rho \nabla_t^2 f \tag{6.13}$$

where f is any differentiable function and ∇_t^2 is the transverse Laplacian operator

$$\nabla_t^2 = \frac{1}{\rho}\frac{\partial}{\partial \rho}\left(\rho \frac{\partial}{\partial \rho}\right) + \frac{1}{\rho^2}\frac{\partial^2}{\partial \phi^2} \tag{6.14}$$

Hence the E-term in (6.11) vanishes, on taking the real part, apart from the effect of the variation of Γ. This variation is due to the presence of h, and it is the χ term in h which carries the variability. Hence the coupling depends on χ, and vanishes when χ is zero. The same conclusion holds for the H-term in (6.12). Thus (6.11) and (6.12) become decoupled for the straight guide, and it is only the curvature that causes the coupling.

6.3 Decoupled solutions

Decoupled equations can be found as at the end of section 2.3 by combining E and H with the multiplier N of (2.23). (We shall use the notation of that section, though N_0 could here be identified with $j\hat{\xi}$.)

$$F_\pm = E \pm N_0 H \tag{6.15}$$

Then (6.11) and (6.12) become

$$-k\rho F_\pm = h \,\mathrm{Re}\, D^*\left[\frac{\rho}{kh \mp i\gamma} D(F_\pm)\right] \tag{6.16}$$

This (decoupled) equation is deceptively simple. It is necessary to carry out the indicated differentiations using (6.13) for $D^*[\rho D(\;)]$, recalling that $h = 1 - \chi\rho\cos\phi$. At this stage it is also preferable to return to E_s and H_s, rather than working with E and H. This can be done by working with a new field combination G defined by

$$G_\pm = E_s \pm N_0 H_s = F_\pm / h \tag{6.17}$$

As in section 2.3 it is convenient to drop temporarily the $\pm$ subscript, restoring it later as needed. Then (6.16), after some manipulation and simplification, can be expressed in the form

$$\nabla_t^2 G + (k^2 + \gamma^2/h^2)G = (\chi/\Gamma^2)\Big\{\chi(k^2 - \gamma^2/h^2)G +$$
$$h(k^2 + 3\gamma^2/h^2)\left[\cos\phi\frac{\partial G}{\partial\rho} - \frac{\sin\phi}{\rho}\frac{\partial G}{\partial\phi}\right] \pm 2\gamma k\left[\sin\phi\frac{\partial G}{\partial\rho} + \frac{\cos\phi}{\rho}\frac{\partial G}{\partial\phi}\right]\Big\} \tag{6.18}$$

This is the wave equation for G, in which h appears on the left only in the term γ^2/h^2. On the right the first term varies as χ^2 and the remainder as χ, apart from the variation of h and of G itself with χ.

A perturbation solution can be sought as a power series in χ, assuming

$$G = [G_0 + \chi G_1 + \chi^2 G_2 + \ldots]\, e^{-\gamma s}$$
$$\gamma^2 = \gamma_0^2(1 + \chi A_1 + \chi^2 A_2 + \ldots) \tag{6.19}$$

with γ_0 the propagation constant for the relevant mode in straight guide.

All of the G_n in the above equation are really $G_{n,\pm}$ according to the $\pm$ sign in (6.18). To obtain a solution it is necessary to substitute (6.19) into (6.18) and equate the coefficients of like powers of χ. This will produce a set of differential equations for the G_n generated by the initial choice G_0. Solutions G_n will be selected free of the earlier forms, so as to constitute a unique set just as was done in chapters 3 and 4 for the twisted and curved rectangular guide. In particular, G_1 will be linearly free of G_0. However, degeneracy, when present, gives rise to mode coupling and it is no longer the case that the first-order coefficient A_1 in (6.19) is necessarily zero.

6.4 The circular electric mode in curved guide

This mode is important because of its use in long distance waveguide communication employing the low attenuation properties of the TE_{01} mode. It is derived from an axial magnetic component

$$H_s = J_0(\Gamma_0\rho)\, e^{-\gamma_0 s} \tag{6.20}$$

with

$$\Gamma_0 = j_{11}/a, \quad J_1(j_{11}) = 0 \tag{6.21}$$

and a is the guide radius. As already discussed this mode is degenerate with the TM_{11} mode derived from an axial electric component.

$$E_s = J_1(\Gamma_0\rho)\sin\phi e^{-\gamma_0 s} \tag{6.22}$$

If either form alone is assumed as a basis G_0 in (6.19) it is found that (6.18) cannot be satisfied unless $\chi = 0$, when the solution is trivial. The reason comes from requirements originating in the χ-term. It is found that it is necessary to work with the linear combination

$$G_{0,\pm} = J_1(\Gamma_0\rho)\sin\phi \pm MJ_0(\Gamma_0\rho) \tag{6.23}$$

where M, although arbitrary at this stage, is uniquely determined later and takes on either of two possible values corresponding to two distinct stable mode forms in the curved guide. For each there is a corresponding value of A_1 and of γ.

Substitute (6.19) and (6.23) into (6.18) and equate the coefficients of like powers of χ. The resulting equations are for the G_1 and are rather cumbersome for the higher-order terms: here we shall calculate G_1 which comes from the coefficient of χ. It is found to satisfy the equation

$$\begin{aligned}(\nabla_t^2 + \Gamma_0^2)G_1 &= \{(k^2 + 3\gamma_0^2)[\Gamma_0 J_0(\Gamma_0\rho)\sin\phi\cos\phi - J_1(\Gamma_0\rho)\times \\ &(\pm M\Gamma_0\cos\phi + \rho^{-1}\sin 2\phi)] \pm 2k\gamma_0[\Gamma_0 J_0(\Gamma_0\rho)\sin^2\phi - J_1(\Gamma_0\rho)\times \\ &(\pm M\Gamma_0\sin\phi - \rho^{-1}\cos 2\phi])\}/\Gamma_0^2 - \gamma_0^2\{A_1[J_1(\Gamma_0\rho)\sin\phi \\ &\pm MJ_0(\Gamma_0\rho)] + 2\rho\cos\phi[J_1(\Gamma_0\phi)\sin\phi \pm MJ_0(\Gamma_0\rho)]\}\end{aligned} \tag{6.24}$$

The homogeneous solution, of course, is satisfied by terms such as $J_m(\Gamma_0\rho)\sin m\phi$ and $J_m(\Gamma_0\rho)\cos m\phi$ for any integral value of m; though to avoid introducing a G_0-like term, the combinations in (6.23) with $m = 0$ and 1 are omitted. Such terms on their own do not satisfy the boundary conditions at $\rho = a$ for $m > 1$, however, and are needed only to aid in satisfying these conditions for the particular solution stemming from the right-hand side of (6.24).

To aid in solving (6.24) for the particular solution we can build up a list of results of the consequence of operating with $(\nabla_t^2 + \Gamma_0^2)$ on a variety of likely functions. Clearly these must involve $J_0(\Gamma_0\rho)$ and $J_1(\Gamma_0\rho)$, $\sin\phi$, $\cos\phi$, $\sin 2\phi$, $\cos 2\phi$, and various powers of ρ. It is found that $\rho^n\cos m\phi J_0(\Gamma_0\rho)$ and $\rho^n\cos m\phi J_1(\Gamma_0\rho)$ are suitable forms and give

$$(\nabla_t^2 + \Gamma_0^2)\rho^n\cos m\phi J_0(\Gamma_0\rho) = \rho^{n-2}\cos m\phi\,[(n^2 - m^2)J_0(\Gamma_0\rho) - 2n\Gamma_0\rho J_1(\Gamma_0\rho)] \tag{6.25}$$

$$(\nabla_t^2 + \Gamma_0^2)\rho^n\cos m\phi J_1(\Gamma_0\rho) = \rho^{n-2}\cos m\phi\,[(n^2 - 2n + 1 - m^2)J_1(\Gamma_0\rho) + 2n\Gamma_0\rho J_0(\Gamma_0\rho)] \tag{6.26}$$

There is a similar pair with $\cos m\phi$ replaced by $\sin m\phi$. These results enable combinations of m and n to be found that reproduce all the terms on the right of (6.24).

The resulting solution is

$$\begin{aligned}G_{1,\pm} = {} & J_0(\Gamma_0\rho)\,[\rho^2 \sin 2\phi(\gamma_0^2\Gamma_0/4) + \rho \sin\phi(k\gamma_0 M + \tfrac{1}{2}\gamma_0^2\Gamma_0 A_1) \\ & \pm \rho\cos\phi\, M(k^2 + 3\gamma_0^2)/2]\,/\Gamma_0^2 - J_1(\Gamma_0\rho)\,[\pm\rho^2\cos\phi \times \\ & (\tfrac{1}{2}M\Gamma_0\gamma_0^2) \pm \rho\cos 2\phi(\gamma_0 k/2) - \rho\sin 2\phi(k^2 + 3\gamma_0^2)/4 \\ & \pm \rho(\Gamma_0\gamma_0^2 M A_1 - k\gamma_0)]\,/\Gamma_0^2 + BJ_2(\Gamma_0\rho)\sin 2\phi \\ & \pm CJ_2(\Gamma_0\rho)\cos 2\phi \pm DJ_1(\Gamma_0\rho)\cos\phi \end{aligned} \qquad (6.27)$$

The terms with arbitrary multiplying constants B, C and D arise as solutions of the homogeneous equation and are needed in satisfying the boundary conditions. Since the latter must hold at $\rho = a$ for all values of ϕ, appropriate terms are needed for cancellation in all cases where vanishing is not ensured by the form of the Bessel function. The $\pm$ on C and D is needed to match the $\pm$ on the corresponding $\cos 2\phi$ and $\cos\phi$ terms arising as solutions of the particular integral.

6.5 Boundary conditions

At $\rho = a$ we need $E_s = 0$. From (6.17) this gives

$$G_{1,+} + G_{1,-} = 0, \quad \rho = a \qquad (6.28)$$

Since $J_1(\Gamma_0\rho) = 0$ at $\rho = a$ from (6.21), (6.28) gives

$$\begin{aligned}& J_0(\Gamma_0 a)\,[\tfrac{1}{2}a^2\gamma_0^2\Gamma_0 \sin 2\phi + a(2M\gamma_0 k + \gamma_0^2\Gamma_0 A_1)\sin\phi] \\ & + 2B\Gamma_0^2 J_2(\Gamma_0 a)\sin 2\phi = 0\end{aligned} \qquad (6.29)$$

The coefficients of both the $\sin 2\phi$ and $\sin\phi$ terms must each be separately zero to satisfy this equation for all ϕ. Using the relation $J_2(x) = -J_0(x) + 2J_1(x)/x = -J_0(x)$ when $x = \Gamma_0 a = j_{11}$ gives

$$\begin{aligned} B &= a^2\gamma_0^2/4\Gamma_0 \\ M &= -A_1\gamma_0\Gamma_0/2k \end{aligned} \qquad (6.30)$$

The latter relation between M and A_1 goes part way to establishing a unique value for M.

The second set of boundary conditions stems from $E_\phi = 0$, which from (6.4) is proportional to $\partial H/\partial\rho$, since H_ρ is zero at $\rho = a$.

Now $H = hH_s$, whence, from (6.17), to order χ^2,

$$0 = h\partial H_s/\partial\rho + H_s\partial h/\partial\rho$$

$$= (1-\chi\rho\cos\phi)\frac{\partial}{\partial\rho}\,[(G_{0,+}-G_{0,-})+\chi(G_{1,+}-G_{1,-})]$$

$$-\chi\cos\phi\,[(G_{0,+}-G_{0,-})+(G_{1,+}-G_{1,-})]\,,\ \rho=a. \qquad (6.31)$$

Now $G_{0,+}-G_{0,-}$, from (6.23), $= 2MJ_0(\Gamma_0\rho)$; its differential coefficient vanishes at $\rho=a$. Hence (6.31), to $0(\chi)$, gives

$$0 = \frac{\partial}{\partial\rho}(G_{1,+}-G_{1,-})-2MJ_0(\Gamma_0\rho)\cos\phi,\ \rho=a. \qquad (6.32)$$

This formula corrects an earlier treatment[3] from which the last term was missing. Its main effect, however, is via the constant D in (6.27); it does not affect the propagation constant or the coupling coefficient. Inserting (6.32) into (6.27) gives

$$J_0(\Gamma_0 a)\,[M(k^2+3\gamma_0^2-2\Gamma_0^2-\gamma_0^2\Gamma_0^2a^2)\cos\phi$$

$$+a\gamma_0\Gamma_0(k-\gamma_0\Gamma_0MA_1)-ak\gamma_0\Gamma_0\cos 2\phi]$$

$$+2D\Gamma_0^3\cos\phi J_1'(\Gamma_0 a)+2C\Gamma_0^3\cos 2\phi J_2'(\Gamma_0 a)=0 \qquad (6.33)$$

Now $J_1'(x)=J_0(x)-J_1(x)/x=J_0(x)$ at $x=j_{11}$, and $J_2'(x)=J_1(x)(1-4/x^2)+2J_0(x)/x=2J_0(x)/x$ at the same value of x. Hence, from the coefficients of $\cos\phi$, $\cos 2\phi$ and the constant term, (6.33) gives

$$\left.\begin{aligned} M\,[k^2+3\gamma_0^2-2\Gamma_0^2-\gamma_0^2\Gamma_0^2a^2]+2D\Gamma_0^3 &= 0\\ C &= a^2k\gamma_0/4\Gamma_0\end{aligned}\right\} \qquad (6.34)$$

$$k = \gamma_0\Gamma_0MA_1 \qquad (6.35)$$

Combining (6.35) and (6.30) for M gives

$$M^2 = -\tfrac{1}{2}, \qquad A_1 = k/\gamma_0\Gamma_0M \qquad (6.36)$$

This relation gives two values of the coupling coefficient: $M_1=j/\sqrt{2}$ and $M_2=-j/\sqrt{2}$, with two corresponding values of A_1. If M stands for either of these two the resulting axial fields take the form

$$E_s = e^{-s(\gamma_0+k\chi/2\Gamma_0M)}\{J_1(\Gamma_0\rho)\sin\phi+ \qquad (6.37)$$

$$(\chi/4\Gamma_0)\sin 2\phi\,[J_0(\Gamma_0\rho)\gamma_0^2(\rho^2-a^2)+J_1(\Gamma_0\rho)(\Gamma_0\rho+2(\rho+a)\gamma_0^2/\Gamma_0)]\}$$

$$\hat{\zeta}H_s = -jMe^{-s(\gamma_0+k\chi/2\Gamma_0M)}\{J_0(\Gamma_0\rho)+$$

$$(\chi/2\Gamma_0)J_0(\Gamma_0\rho)\,[\Gamma_0\rho(1+2\gamma_0^2/\Gamma_0^2)\cos\phi+a^2\gamma_0kM\cos 2\phi]+$$

$$(\chi/2\Gamma_0)J_1(\Gamma_0\rho)\,[\cos 2\phi(\rho^2-a^2)2M\gamma_0k/\Gamma_0\rho-$$

$$\cos\phi(\gamma_0^2(\rho^2-a^2)-1+2\gamma_0^2/\Gamma_0^2)]\} \qquad (6.38)$$

6.6 Mode transformation

The two values $\pm j/\sqrt{2}$ for M in (6.37) and (6.38) correspond to two distinct field forms, propagating with different velocities. (The $\pm$ signs which enter here have nothing to do with the $\pm$ forms used as a means of dealing with the decoupled equations; they merely distinguish between the two values M_1 and M_2 of M.) The difference in the propagation constants is $j\chi\sqrt{2}k/\Gamma_0 = jk\sqrt{2}/\Gamma_0 R$, so that these modes are in anti-phase in a length $s_c = \sqrt{2}\lambda\Gamma_0 R/4$, corresponding to an angle of bend $\theta_c = s_c/R = \sqrt{2}j_{11}\lambda/4a$. To this order the angle is independent of the radius of the bend though it is proportional to the wavelength. It is known as the critical angle, and is that bend angle in which an original TE_{01} mode is completely transformed into the TM_{11} structure. To demonstrate this, suppose a pure TE_{01} mode is incident at the entrance of a curved section of pipe. It is obtainable from an axial field H_s given by (6.20). Now if $(E_s, H_s)_{\pm}$ represents the combinations of (6.23), which are the stable modes in the curved pipe, then we can put

$$H_{s(TE_{01})} = [(E_s, H_s)_+ - (E_s, H_s)_-]/\sqrt{2}, \quad s = 0. \qquad (6.39)$$

The two mode combinations in this equation start off in phase so that at $s = 0$ they reduce to the initial TE_{01} form. After transversing a distance s_c the modes are in anti-phase and the composite mode is obtainable from an axial field of approximately $e^{-\gamma_0 s_c}[(E_s, H_s)_+ + (E_s, H_s)_-]/\sqrt{2} = \sqrt{2}E_s e^{-\gamma_0 s_c}$. This is the form of the TM_{11} mode, and indicates how the initial mode is transformed by the bend. After a further distance s_c the mode reverts back to TE_{01}, and can be recovered in this form if a transition to straight guide occurs at this position. Unfortunately the angle through which this process takes place is frequency dependent, since θ_c is proportional to λ.

This form of description was first given by Jouguet[1], and was originally suggested as a means of negotiating bends: though because of the frequency dependence of θ_c the method is not a broad-band one. More recent solutions use a surface reactance in the form of a circumferential or helical disturbance to the guide surface; this affects the TM_{11} mode much more than the TE_{01} mode, and therefore breaks the degeneracy. Another way of achieving the same end is to coat the guide wall with a thin coating of low-loss dielectric material.

6.7 The TM_{11} mode in curved guide: first-order solution

As shown in the previous sections, the TM_{11} mode with electric polarization perpendicular to the plane of the bend couples with a unique amount of TE_{01} mode to constitute a stable mode of the curved guide, i.e. a mode structure that is approximately maintained in the curved guide, and to which perturbation techniques can be applied. In contrast, the TM_{11} mode with electric polarization parallel to the plane of the bend is already such a natural mode and perturbation methods apply directly. We start with (6.18) and (6.19) and chose

$$G_0 = J_1(\Gamma_0\rho)\cos\phi \tag{6.40}$$

with $\Gamma_0 = j_{11}/a$ as before.

Analogously to (6.27) one finds

$$\begin{aligned}(\nabla_t^2 + \Gamma_0^2)G_{1,\pm} &= J_0(\Gamma_0\rho)[(k^2 + 3\gamma_0^2)\cos^2\phi \pm \gamma_0 k \sin 2\phi]/\Gamma_0 \\ &-J_1(\Gamma_0\rho)\{[(k^2 + 3\gamma_0^2)\cos 2\phi \pm 2\gamma_0 k \sin 2\phi]/\rho\Gamma_0^2 \\ &+ \gamma_0^2(A_1 \cos\phi + 2\rho\cos^2\phi)\}\end{aligned} \tag{6.41}$$

To find the particular solution we use the results of (6.25) and (6.26) to give

$$\begin{aligned}G_{1,\pm} &= \rho J_0(\Gamma_0\rho)\gamma_0^2[\rho(1 + \cos 2\phi) + A_1 \cos\phi]/4\Gamma_0 \\ &+ \rho J_1(\Gamma_0\rho)[1 + (1 + 2\gamma_0^2/\Gamma_0^2)\cos 2\phi \\ &\pm (2\gamma_0 k/\Gamma_0^2)\sin 2\phi]/4 + BJ_0(\Gamma_0\rho) \\ &+ CJ_2(\Gamma_0\rho)\cos 2\phi \pm DJ_2(\Gamma_0\rho)\sin 2\phi\end{aligned} \tag{6.42}$$

Here, as in (6.27), solutions of the homogeneous equation are added as appropriate to provide terms suitable for matching the boundary conditions.

From $E_s = 0$ at $\rho = a$ we get, since $G_0 = 0$ at the guide wall,

$$G_{1,+} + G_{1,-} = 0, \quad \rho = a \tag{6.43}$$

Since $J_1(\Gamma_0\rho) = 0$ on the boundary, we get, from the coefficients of the constant term, $\cos 2\phi$ and $\cos\phi$,

$$B = -C = -a^2\gamma_0^2/4\Gamma_0, \quad A_1 = 0 \tag{6.44}$$

The relation $A_1 = 0$ shows that there is no first-order change in the propagation constant, as might have been anticipated.

The boundary condition on E_ϕ requires $\partial h(G_+ - G_-)/\partial\rho = 0$ at $\rho = a$, leading to an equation analagous to (6.32),

$$\partial G_{1,+}/\partial\rho - \partial G_{1,-}/\partial\rho = 0 \tag{6.45}$$

The reason that there are not to this order any additional terms is that, for the initial choice (6.40) for G_0, we have $G_{0+} - G_{0-} = 0$, i.e. there is no initial axial component of magnetic field H_s to contribute. Correct to first order in χ the resulting fields accordingly are

$$E_s = J_1(\Gamma_0\rho)\cos\phi e^{-s\gamma_0} + (\chi/2\Gamma_0)\{\gamma_0^2(\rho^2 - a^2)J_0(\Gamma_0\rho)\cos^2\phi + J_1(\Gamma_0\rho)\Gamma_0\rho\,[\cos^2\phi + (1 + a^2/\rho^2)\cos 2\phi\gamma_0^2/\Gamma_0^2]\}e^{-s\gamma_0} \tag{6.46}$$

$$\hat{\xi}H_s = \chi(\gamma_0 k/4j\Gamma_0^2)\,[2\rho J_1(\Gamma_0\rho) - a^2\Gamma_0 J_2(\Gamma_0\rho)]\sin 2\phi\, e^{-s\gamma_0} \tag{6.47}$$

Higher order solutions can be found in the same way, but the equations rapidly become extremely complicated and cumbersome. To find the effect of the curvature on the propagation constant for instance, requires the term A_2 in (6.19), which in turn leads to the need to investigate G_2 and other terms to order χ^2.

References

1 Jouguet, M.: 'Effects of curvature on the propagation of electromagnetic waves in guides of circular cross-section', *Câbles et Transmission* (Paris), Vol. 1, No. 2, pp. 133–153, July 1947

2 Karbowiak, A.E.: 'Theory of imperfect waveguides: the effect of wall impedance', *Proc. IEE.*, Part B, Vol. 102, No. 5, pp. 698–708, Sept. 1955

3 Lewin, L.: 'Theory of waveguides' (Newnes-Butterworths, London, 1975) p. 109

Chapter 7

Radiation conditions at the boundary of curved structures

7.1 The need to express the radiation condition in local form

One method of treating radiation from curved structures is to attempt a perturbation analysis, treating the curvature as a small perturbation to the straight configuration. In so doing one is immediately concerned with matching fields at the structure surface, and hence with the form of the radiation condition to be applied there. The difficulty arises from the fact that the perturbation analysis presents the field in local co-ordinate form and the range for which it is valid is limited to dimensions of the order of the size of the structure, in the present case the bending radius. Since the radiation condition has to be applied at 'infinity', i.e. at a distance much greater than the bending radius, the local co-ordinate expansion is quite useless. Instead, the local form of the radiation condition needs to be inferred, by an indirect process, and then applied at the surface. It should be stressed that an 'outward-looking' wave in the local co-ordinates is not necessarily (and, in general, is not) the appropriate form to satisfy the radiation condition at very large distances. Thus the out-going Hankel function $H_\nu^{(2)}(k\rho)$ behaves like an outgoing wave when $k\rho \gg \nu$, but looks more like a sum of a (very small) growing and an evanescent wave when $k\rho$ is small. To require only the evanescent term, the usual choice, would be erroneous.

Since in cylindrical co-ordinates the solution can, in any case, be set up rigorously, it might be asked what purpose is served by first obtaining this solution, then using it to determine the local form of the radiation condition, and finally applying this condition to a perturbation analysis. The advantage is threefold. It gives a better insight into the physics of the situation, and may therefore indicate ways of controlling the radiation.[1] Moreover, to get an analytic solution from the rigorous form of solution, asymptotic methods must in any case be applied, and this is readily done directly in the perturbation solution whose development depends on just such an expansion. And finally it points the way to treating problems in other coordinate systems for which no rigorous form of solution exists.

7.2 Poynting vector methods

In the case of the curved dielectric slab it can be readily verified that, in the absence of the growing wave component of field outside the surface a perturbation solution will give no attenuating contribution to the propagation constant. Nevertheless it may be possible to deduce by a different method the correct amount of radiation, and hence to conclude what the attenuation due to bending should be. The situation is somewhat analogous to the use of a sinusoidal approximation to the current on a linear antenna, for which the examination of the corresponding input impedance leads to the conclusion that the arrangement is non-radiating. Nevertheless, by utilizing the expression for the Poynting vector at infinity, a sensible first-order approximation to the radiation can still be obtained. The reason this method works seems to be due to the fact that the expression for the radiation via the Poynting vector is in variational form, and is therefore insensitive to small differences in the current distribution; though these are, however, vital for calculating the real part of the input impedance. Something of the same process was used in Chapter 5 to calculate the second-order junction input impedance from a first-order field form which would, to that order, have indicated no junction effect if calculated directly from the field. Marcuse[2] has performed the equivalent calculation for the curved dielectric slab. Using only the evanescent part of the local field he relates it to the field from a straight slab, whose properties are well-known. Then he takes the field at large distances, calculates the Poynting vector radiation, and deduces the radiation from the curved slab. Finally the attenuation is deduced from the radiation and power flow. This ingenious method does not require the (very small) growing component of the local field, and hence ignores, to a first order, the radiation condition. Snyder *et al.*[3] have performed a similar calculation for the bent fiber by replacing the effect of the fiber core by fictitious electric and magnetic surface currents and using standard antenna theory to calculate the radiation. An example of this process to estimate the radiation at the junction between a straight and curved section of dielectric slab is discussed in Chapter 8.

Valuable though these methods are, they yield neither phase information nor knowledge of the effect of the curvature on the field. Both of these are needed for a thorough discussion of junction effects. We shall accordingly concentrate the development here on the use of the radiation condition combined with perturbation methods.

7.3 The radiation condition in cylindrical co-ordinates

Fig. 7.1 shows a surface bent to a radius of curvature R_0. In a cylindrical co-ordinate system, travelling wave solutions vary as $\exp(-j\nu\phi) = \exp(-jk's)$ where k' is the propagation constant along the surface. Since $s = \phi R_0$ this gives

$$\nu = k'R_0 \tag{7.1}$$

as the connection between the azimuthal index ν, the propagation constant and the radius of the curvature.

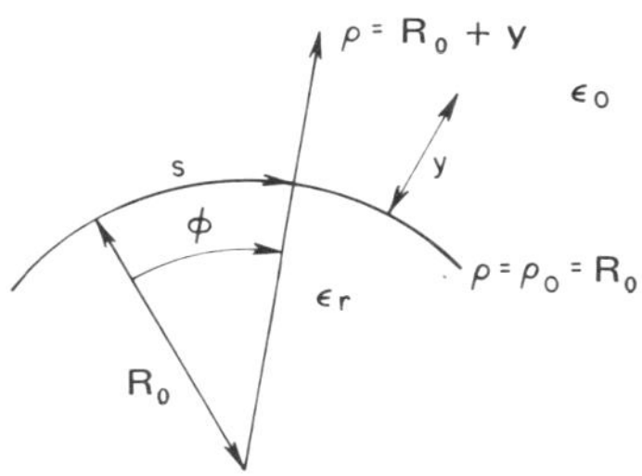

Fig. 7.1 Co-ordinates for curved surface

It is important to assess the order of magnitude of ν. The applications of interest here are to dielectric materials of relative permittivity ϵ_r, so that k' may be expected to lie between k_0 and $k_1 = k_0\epsilon_r^{1/2}$, with k_0 the propagation constant of the space outside the dielectric, assumed here to be free space.

In cylindrical co-ordinates, outgoing solutions for $\rho > R_0$ are built up from forms such as $\exp(-j\nu\phi)H_\nu^{(2)}(k_0\rho)$. At the surface, $\rho = \rho_0 = R_0$, so that $k_0\rho_0/\nu = k_0R_0/k'R_0 = k_0/k'$. This is a number of the order of, but less than, unity. Since R_0, and hence ν will be large for small curvatures, we shall accordingly be concerned with asymptotic expansions of the Hankel function for large order and large argument.

The region $k_0\rho/\nu < 1$ is adjacent to the surface and defines a local region in which the argument of the Hankel function is less than the order. If we put $\rho = R_0 + y$, y is a co-ordinate that increases from zero above the surface, and is the local co-ordinate in which the field quantities can be expressed. The restriction $k_0\rho < \nu$ sets an upper limit to the value of y for which formulas based on the requirements of this restriction are valid. This makes it quite clear how a wave form growing with y could be permissible without violating the radiation condition, which is applied for $\rho \to \infty$. The second type Hankel function automatically satisfies the radiation condition for large ρ; the local form is to be found by investigating the Hankel function structure for $\rho = R_0 + y$, with $y \ll R_0$ and $k_0\rho < \nu$.

Debye's asymptotic expansions[4] can be put in the form

$$(2\pi\nu)^{1/2} J_\nu(\nu z) \sim (1 - z^2)^{-1/4} \exp\left[-\nu f(z)\right] \tag{7.2}$$

$$(2\pi\nu)^{1/2} Y_\nu(\nu z) \sim -2(1 - z^2)^{-1/4} \exp\left[+\nu f(z)\right] \tag{7.3}$$

where

$$f(z) = \tanh^{-1}(1 - z^2)^{1/2} - (1 - z^2)^{1/2} \tag{7.4}$$

and z is less than unity.

If we take $\nu z = k_0(R_0 + y)$, with $y \ll R_0$, we can expand $f(z)$ to first order about $z_0 = k_0 R_0/\nu$ and obtain

$$\nu f(z) = \nu f(k_0 R_0/\nu) - y(k'^2 - k_0^2)^{1/2}, \quad y \ll R_0. \tag{7.5}$$

Since $H_\nu^{(2)} = J_\nu - jY_\nu$ we can combine (7.2) and (7.3) to find the local form of the field for small y. From (7.3) alone we would have an exponentially decreasing term, but (7.2) is exponentially increasing, and the two terms together appear in the relative combination

$$\exp\left[-y(k'^2 - k_0^2)^{1/2}\right] - \tfrac{1}{2}j \exp\left[+y(k'^2 - k_0^2)^{1/2} - 2\nu f(k_0 R_0/\nu)\right] \tag{7.6}$$

The second term in (7.6) gives the modification to the expected decaying first term to account for the local form taken by the radiation condition. It is normally of very small magnitude because of the presence of the factor $\exp\left[-2\nu f(k_0 R_0/\nu)\right]$, but because of the initial factor j it is in quadrature with the first term and it is this that leads eventually to the radiation.

Below the surface the field has to be finite at $\rho = 0$, and hence $H_\nu^{(2)}$ must be replaced by J_ν. Since y decreases below the surface, (7.2), along with (7.5), indicates a simple decaying wave; in this region the expected decaying component suffices. This requirement, together with (7.6) above the surface, is sufficient for enabling field matching to be achieved at the surfaces of a curved dielectric slab. The analysis of this configuration is discussed in Chapter 8.

An alternative approach to this aspect of the radiation condition is given in section 7.7 where the growing component of the wave is interpreted as the reflected component of the decaying wave, the 'reflection' occurring at the boundary $k\rho = \nu$ where the character of the Debye asymptotic expansion changes. This interpretation enables some physical insight into the process to be attained, from which measures to reduce the radiation can be taken.[1]

7.4 The radiation condition for a toroidal structure: initial considerations

Fig. 7.2 shows the toroidal geometry upon which three different co-ordinate systems have been erected. Spherical co-ordinates, R, θ, ϕ, are used in the investigation of the behavior at infinity. Cylindrical co-ordinates r, ϕ, z are used to construct an exact and suitable solution to the Helmholtz equation. Local co-ordinates ρ, ψ, s approximate to cylindrical co-ordinates on the torus when the torus radius R_0 is very large. Note that the torus axial co-ordinate is $s = R_0\phi$, so that the relevant axial fields in the local co-ordinate system are E_ϕ and H_ϕ.

The co-ordinates are inter-related by

$$\begin{aligned} z &= R\cos\theta = -\rho\sin\psi \\ r &= R\sin\theta = R_0 + \rho\cos\psi \end{aligned} \qquad (7.7)$$

We define

k = propagation coefficient in the space outside the torus
k' = propagation coefficient around the torus ($k' > k$ for a wave on a dielectric rod embedded in a medium of lower permittivity)
$\gamma = k'/k > 1$
$\Gamma = (1 - 1/\gamma^2)^{1/2}$
$(\gamma^2 - 1)^{1/2} = |\gamma^2 - 1|^{1/2} e^{-j\delta}, \delta > 0$ if the torus dielectric is lossy
$\beta = \sinh^{-1}(\gamma^2 - 1)^{-1/2}$

The quantities β, γ and Γ will be needed in due course. It is convenient to consider the dielectric always slightly lossy so that $\delta > 0$. Then β will be complex, $= \beta_r + j\beta_i$, say. From $\sinh\beta = (\gamma^2 - 1)^{-1/2}$ we get $\tan\beta_i = \tan\delta \tanh\beta_r$, whence $\beta_i < \delta$ since $\tanh\beta_r < 1$.

The space can be divided into the two regions $r < R_0$ and $r > R_0$. In the inner region the field must be finite at $r = 0$. In the outer region the field should look like an outgoing wave as $r \to \infty$. Outward-going waves in the $\pm z$ directions imply sources in the plane $z = 0$, so that forms like $\exp(-j\mu|z|)$, with μ a suitable parameter, will be involved in setting up solutions. The outer region corresponds to $-\pi/2 < \psi < \pi/2$ but if solutions based on $\exp(-j\mu|z|)$ are used only the range $0 < \psi < \pi/2$ need be considered since the use of $|z|$ automatically looks after the range $-\pi/2 < \psi < 0$.

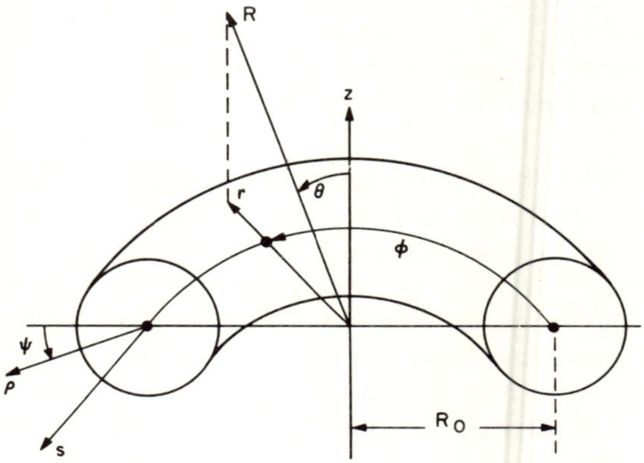

Fig. 7.2 Toroidal co-ordinate system

The aim is to develop a field which satisfies Maxwell's equations, behaves correctly at infinity and at the origin, and which can be developed on the torus to give a dominant term corresponding to an outgoing or attenuated wave in the local cylindrical co-ordinate system. The form of such a wave is given by $\exp(-jk's)K_m[k\rho(\gamma^2 - 1)^{1/2}]$ $\cos m\psi$, this being the symmetrical form of solution for a straight cylinder. The anti-symmetrical solution has $\cos m\psi$ replaced by $\sin m\psi$. Any additional terms found to the above form will be interpreted as the local structure of the radiation condition, in much the same way as was done for the extra term in (7.6).

An elementary solution of the Helmholtz equation for E_ϕ in the basic cylindrical system r, ϕ, z can be put in the form

$$E_\phi = \exp[jk\mu z - j\nu\phi]\, K'_\nu[kr(\mu^2 - 1)^{1/2}] \qquad z < 0, r > R_0$$

$$E_\phi = \exp[jk\mu z - j\nu\phi]\, I'_\nu[kr(\mu^2 - 1)^{1/2}] \qquad z < 0, r < R_0 \tag{7.8}$$

with $\nu = k'R_0$. This clearly corresponds to a variation $\exp(-jk's)$ around the torus and is suitable for building up a general solution by multiplying by a weighting function of μ and integrating with respect to μ over a suitable contour. The problem is to choose a suitable weighting function and contour so as to give the form leading to the dominant term of an outgoing wave in the local coordinate system, and then to deduce the radiation correction to this term from the integral representation.

7.5 The range $r > R_0$

In (7.8) take μ in the form $(\gamma^2 - 1)^{1/2} \sinh x$, with x a variable of integration. Replace z by $-\rho \sin \psi$, r by $R_0 + \rho \cos \psi$, $0 < \psi < \pi/2$, multiply by a weighting function $F(x)$ and integrate over a suitable contour

$$E_\phi = \exp(-j\nu\phi) \int F(x) \exp[-jk\rho(\gamma^2 - 1)^{1/2} \sin\psi \sinh x]\, K_\nu'[k(R_0 + \rho\cos\psi)\Gamma_x]\, dx \tag{7.9}$$

with

$$\Gamma_x = [(\gamma^2 - 1)\sinh^2 x - 1]^{1/2}$$

At $\rho = 0$ this equation gives the field for $z < 0$ on the cylinder $r = R_0$. If this field had been represented by a Fourier integral between infinite limits for μ, the corresponding limits for x would be $\pm(\infty + j\delta)$, since the factor $(\gamma^2 - 1)^{1/2}$ brings in a phase angle $-\delta$. Accordingly we take x along the contour of Fig. 7.3*a*. The branch cuts at $\pm\beta$ come from the form of Γ_x in (7.9), since β is defined to give $\Gamma_x = 0$ at $x = \beta$. The contour goes over the cuts in the way shown since $\beta_i < \delta$, as indicated in section 7.4.

The demonstration that (7.9) does indeed give an outgoing wave at infinity involves transforming to spherical co-ordinates via (7.7) and taking a new integration variable y defined by

$$(\gamma^2 - 1)^{1/2} \sinh x = -j \sinh y \tag{7.10}$$

Near the origin this gives $y \sim j(\gamma^2 - 1)^{1/2} x$. Moreover, since $\arg(\gamma^2 - 1)^{1/2} = -\delta$ we get $y = \pm(-\infty + j\pi/2)$ at $x = \pm(\infty + j\delta)$. The contour for y is therefore as shown in Fig. 7.3*b*. The branch cuts rotate to $y = \pm j\pi/2$ and are not encountered. Equation (7.9) becomes

$$E_\phi = \exp(-j\nu\phi) \int_{-\infty + j\pi/2}^{\infty - j\pi/2} \exp(kR\cos\theta \sinh y) H_\nu^{(2)\prime}(kR \sin\theta \cosh y)\, F(x)(dx/dy)dy \tag{7.11}$$

For large R we can approximate the Hankel function by

$$\exp\{-j[kR\sin\theta\cosh y - \nu\pi/2 - \pi/4]\}(2/\pi kR\sin\theta\cosh y)^{1/2}.$$

The variable part of the total exponent in the integrand is therefore

$$-jkR(\cosh y\sin\theta + j\sinh y\cos\theta) = -jkR\cosh(y - j\theta + j\pi/2).$$

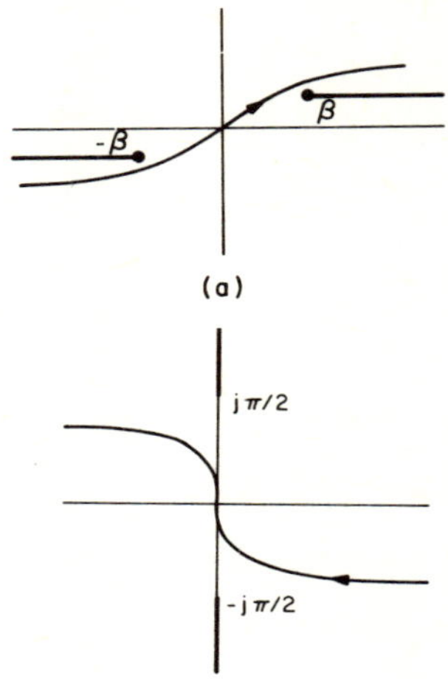

Fig. 7.3 *a* Contour of integration for *x*
b Contour of integration for *y*

Since $\pi/2 < \theta < \pi$ in the region considered ($z < 0$) we can move the y-contour to run horizontally through the point $j(\theta - \pi/2)$, and on writing $y' = y - j(\theta - \pi/2)$ we get the modified exponent $-jkR \cosh y'$. For large R the method of stationary phase gives its contribution from integration through $y' = 0$, and with the factor $R^{-1/2}$ already derived from the Hankel function we get a total variation of the form e^{-jkR}/R for the integral. This verifies the outward-going character of (7.9) at infinity. It remains to find $F(x)$ to give the required form at the torus. To do this we require some results on Bessel functions whose order and argument are of the same order. Results using Debye's expansions[4] are developed in Appendix C, and quoting from there we have, for large ν and finite b and z,

$$\frac{K_\nu'(\nu z + bz)}{K_\nu'(\nu z)} =$$

$$\exp[-b(1+z^2)^{1/2}]\left\{1 + \frac{b}{2\nu}\left[\frac{1}{(1+z^2)^{1/2}} - \frac{b(2+z^2)}{1+z^2}\right] + 0(\nu^{-2})\right\} \tag{7.12}$$

The term in ν^{-1} is finite for real z and approaches zero as $\nu \to \infty$. Since we will be interested in results only for large ν in this section this term will be ignored, and we can put, for $\nu \gg 1$,

$$K'_\nu(\nu z + bz)/K'_\nu(\nu z) \sim \exp\,[-b(1+z^2)^{1/2}] \tag{7.13}$$

In a similar way we get

$$J'_\nu(\nu y + by)/J'_\nu(\nu y) \sim \exp\,[b(1-y^2)^{1/2}], \qquad y^2 < 1 \tag{7.14}$$

$$Y'_\nu(\nu y + by)/Y'_\nu(\nu y) \sim \exp\,[-b(1-y^2)^{1/2}], \quad y^2 < 1 \tag{7.15}$$

$$J'_\nu(\nu y)/Y'_\nu(\nu y) \sim \tfrac{1}{2}\exp\,[-2\bar{f}(y)] \tag{7.16}$$

where

$$\bar{f}(y) = \tanh^{-1}(1-y^2)^{1/2} - (1-y^2)^{1/2} \tag{7.17}$$

and finally we have

$$\begin{aligned} &K'_\nu(\nu y) I_\nu(\nu z) \sim -1/2\nu z \\ &Y'_\nu(\nu y) J_\nu(\nu y) \sim 1/\pi\nu y, \quad 0 < y < 1 \end{aligned} \tag{7.18}$$

Combining (7.12) and (7.18) we see that, for z real

$$-\nu z I_\nu(\nu z) K'_\nu(\nu z + bz) \sim \tfrac{1}{2}\exp\,[-b(1+z^2)^{1/2}] \tag{7.19}$$

But for z imaginary, $= jy$, we use the results

$$I_\nu(ju) = e^{j\nu\pi/2} J_\nu(u) \tag{7.20}$$

and

$$K'_\nu(ju) = -\tfrac{1}{2}\pi e^{-j\nu\pi/2} H_\nu^{(2)\prime}(u) = -\tfrac{1}{2}\pi e^{-j\nu\pi/2}\,[J'_\nu(u) - jY'_\nu(u)] \tag{7.21}$$

to get

$$\begin{aligned} -\nu z I_\nu(\nu z) K'_\nu(\nu z + bz) &= \tfrac{1}{2}\{\exp\,[-b(1-y^2)^{1/2}] \\ &+ \tfrac{1}{2}j\exp\,[-2\nu\bar{f}(y) + b(1-y^2)^{1/2}]\} \end{aligned} \tag{7.22}$$

with $\bar{f}(y)$ given by (7.17).

It will be noticed that the first term on the right of (7.22) is just the continuation of the function on the right of (7.19) for imaginary z. The second term in (7.22) is *additional* and only appears in that part of the range of z for which z is imaginary. As we shall see, its effect is to give inward or growing waves in the local co-ordinate system.

In order to align (7.22) with the structure of (7.9) we can take $\nu z = kR_0\Gamma_x$, $bz = \rho\cos\psi\Gamma_x$, so that z is real for $x > \beta$ or $< -\beta$, and is imaginary for $-\beta < x < \beta$. It is also convenient to redefine $F(x)$ to incorporate the factor $-\nu z I_\nu(\nu z)$ by putting

$$F(x) = -kR_0\Gamma_x I_\nu(kR_0\Gamma_x)G(x) \tag{7.23}$$

The integral in (7.9) breaks up into two parts E_1 and E_2. E_1 comes from the integration of the form in (7.21) over the whole range; E_2 comes from the integration of the second term in (7.22), and is present only for $-\beta < x < \beta$. Hence, for large R_0, we get $E_\phi = E_1 + E_2$ where

$$E_1 = \tfrac{1}{2}e^{-j\nu\phi}\int_{-\infty-j\delta}^{\infty+j\delta} G(x)\exp[-k\rho(\gamma^2-1)^{1/2}$$
$$(\cosh x\cos\psi + j\sinh x\sin\psi)]\,dx \tag{7.24}$$

$$E_2 = (j/4)e^{-j\nu\phi}\int_{-\beta}^{\beta} G(x)e^{-2\nu f_x}\exp[k\rho(\gamma^2-1)^{1/2}$$
$$(\cosh x\cos\phi - j\sinh x\sin\psi)]\,dx \tag{7.25}$$

with

$$f_x = \tanh^{-1}(\Gamma\cosh x) - \Gamma\cosh x,\ \Gamma = (1-1/\gamma^2)^{1/2} \tag{7.26}$$

The term in $e^{-2\nu f_x}$ goes exponentially to zero as $\nu\to\infty$, and it is only the fact that E_2 is in quadrature to E_1 that makes its presence important. Clearly E_1 is the part of E_ϕ which has to reduce to the field $e^{-j\nu\phi}K_m[k\rho(\gamma^2-1)^{1/2}]\,e^{\mp jm\psi}$ as $R_0\to\infty$, and $G(x)$ needs to be chosen to achieve this. We show now that $G(x) = e^{\pm mx}$ is the required solution. Putting this into (7.24), and noting that the argument of the exponential there can be re-written in terms of $\cosh(x+j\psi)$, we can take $x + j\psi = u$ and get

$$E_1 = \tfrac{1}{2}e^{-j\nu\phi}\int_{-\infty+j(\psi-\delta)}^{\infty+j(\psi+\delta)} e^{\pm m(u-j\psi)}\exp[-k\rho(\gamma^2-1)^{1/2}\cosh u]\,du \tag{7.27}$$

Since $\arg(\gamma^2-1)^{1/2} = -\delta$, the integrand is convergent at both limits for $0 < \psi < \pi/2$. Moreover (7.27) has no singularities so the contour can be displaced to the real u-axis. Hence,

$$E_1 = \tfrac{1}{2}e^{-j\nu\phi}\int_{-\infty}^{\infty} e^{\pm mu}e^{\mp jm\psi}\exp[-k\rho(\gamma^2-1)^{1/2}\cosh u]\,du \tag{7.28}$$

Now $e^{\pm mu} = \cosh mu \pm \sinh mu$. Since the rest of the integrand is even in u the sinh mu term integrates to zero, whence

$$E_1 = e^{-j\nu\phi}\int_0^{\infty} e^{\mp jm\psi}\cosh mu\exp[-k\rho(\gamma^2-1)^{1/2}\cosh u]\,du$$
$$= e^{-j\nu\phi}e^{\mp jm\psi}K_m[k\rho(\gamma^2-1)^{1/2}] \tag{7.29}$$

By combining results for $+m$ and $-m$ we see that to get a variation $\cos m\psi$, $G(x)$ would need to be $\cosh mx$; while $j \sinh mx$ gives a variation $\sin m\psi$.

The term giving rise to the radiation condition is E_2 in (7.25). Taking first $G(x) = \cosh mx$ we note that the integrand is completely dominated by the factor $\exp(-2\nu f_x)$ for ν large. This factor has its greatest value when f_x is smallest, i.e. at $x = 0$. Elsewhere it decreases rapidly, to zero at the limits. Hence a saddle point calculation around $x = 0$ is sufficient and yields, on expanding the remaining exponent as a sum of modified Bessel functions,

$$E_2 \sim (j/4)e^{-j\nu\phi} \sum_0^\infty \epsilon_n I_n [k\rho(\gamma^2 - 1)^{1/2}] \cos n\psi \exp[-2\nu(\tanh^{-1}\Gamma - \Gamma)] .$$
$$\int_{-\infty}^{\infty} \cosh mx \exp[-\nu x^2 \Gamma^3/(1-\Gamma^2)]\, dx$$
$$\sim (j/4)e^{-j\nu\phi} \sum_0^\infty \epsilon_n I_n [k\rho(\gamma^2 - 1)^{1/2}] \cos n\psi$$
$$\exp[-2\nu(\tanh^{-1}\Gamma - \Gamma)]\, (\pi/kR_0)^{1/2} (\gamma^2 - 1)^{-3/4} \qquad (7.30)$$

where
$$\epsilon_0 = 1, \epsilon_n = 2 \text{ for } n > 0.$$

If instead of $\cosh mx$ we had used $j \sinh mx$ the process would have given $E_2 = 0$ to this order for the corresponding integral.

We define the (small) quantity σ by the equation

$$\sigma = (j/4) \exp\{-2kR_0 [\gamma \tanh^{-1}(1 - 1/\gamma^2)^{1/2} - (\gamma^2 - 1)^{1/2}]\}$$
$$(\pi/kR_0)^{1/2} (\gamma^2 - 1)^{-3/4} \qquad (7.31)$$

Then the local form of the field component E_ϕ (the same holds for H_ϕ) is

$$(1)\ e^{-j\nu\phi} K_m [k\rho(\gamma^2 - 1)^{1/2}] \sin m\psi + \sigma 0(\nu^{-1})$$
$$\qquad (7.32)$$
$$(2)\ e^{-j\nu\phi} \{K_m [k\rho(\gamma^2 - 1)^{1/2}] \cos m\psi +$$
$$+ \sigma \sum_0^\infty \epsilon_n I_n [k\rho(\gamma^2 - 1)^{1/2}] \cos n\psi\}$$

The radiation condition therefore gives, to this order of approximation, no additional terms for modes antisymmetrical in ψ. For modes symmetrical about $\psi = 0$ it is seen that the radiation condition involves not only a (growing) I_m term, of magnitude $\sigma\epsilon_m$, but also a train of coupled angular modes; in much the same way as a propagating mode in a periodic structure is accompanied by a set of spatial harmonics, all propagating together with the phase of the main mode.

7.6 The range $r < R_0$

The results of the previous section were limited to $0 < \psi < \pi/2$, though the extension to $-\pi/2 < \psi < 0$ follows by symmetry from the initial choice of $\exp(-j\mu|z|)$ to represent outgoing waves for both positive and negative z. Inside the cylinder $r = R_0$ we require a form that is finite at the origin and gives E_ϕ continuous at $\psi = \pi/2$.

If we take the form of (7.9) with $F(x) = -kR_0\Gamma_x I_\nu(kR_0\Gamma_x)e^{\pm mx}$, replace K_ν' by I_ν' for finiteness at the origin, and add a further factor $K_\nu'(kR_0\Gamma_x)/I_\nu'(kR_0\Gamma_x)$ then the integrands become identical at $\psi = \pi/2$. Hence we are led to consider the form

$$ {}_1E_\phi = -e^{-j\nu\phi}\int_{-\infty-j\delta}^{\infty+j\delta} F_1(x)\exp\left[-jk\rho(\gamma^2-1)^{1/2}\sin\psi\right. $$

$$ \left.\sinh x\right] I_\nu'[k(R_0+\rho\cos\psi)\Gamma_x]\,dx \qquad (7.33) $$

with

$$ F_1(x) = kR_0\Gamma_x e^{\pm mx} I_\nu(kR_0\Gamma_x)K_\nu'(kR_0\Gamma_x)/I_\nu'(kR_0\Gamma_x) \qquad (7.34) $$

Analogously to (7.27) we get

$$ {}_1E_\phi = \tfrac{1}{2}e^{-j\nu\phi}\int_{-\infty+j(\pi-\psi-\delta)}^{\infty+j(\pi-\psi+\delta)} \exp[\pm m(u-j\pi+j\psi)] $$

$$ \exp[-k\rho(\gamma^2-1)^{1/2}\cosh u]\,du \qquad (7.35) $$

Since, now, $(\pi-\psi)$ lies between 0 and $\pi/2$ the same considerations that led to (7.29) now give

$$ {}_1E_\phi = e^{-j\nu\phi}e^{\mp jm(\pi-\psi)}K_m[k\rho(\gamma^2-1)^{1/2}] \qquad (7.36) $$

Hence to get a variation $\cos m\psi$ we need a factor in $F_1(x)$ of $\cos m\pi \cosh mx$, and to get a variation $\sin m\psi$ we need a factor $-j\cos m\pi \sinh mx$. Since the corresponding factors in $F(x)$ were $\cosh mx$ and $j \sinh mx$ it might appear that incorporating the extra factors $\pm\cos m\pi$ would destroy the continuity at $\psi = \pi/2$ on which (7.33) is based. However, $\cos m\psi = 0$, when m is odd, at $\psi = \pi/2$, so that the continuity is maintained irrespective of any initial factor; and the same goes for $\sin m\psi$ when m is even. Hence, instead of (7.34) we need

$$F_2(x) = kR_0\Gamma_x \cos m\pi \ \ e^{\mp mx} I_\nu(kR_0\Gamma_x) K'_\nu(kR_0\Gamma_x)/I'_\nu(kR_0\Gamma_x) \tag{7.37}$$

This form now guarantees that (7.29), which is the dominant part of the field, is obtained over the entire range of ψ.

In addition to the dominant term there is also the radiation term corresponding to (7.25). The part symmetrical about $\psi = 0$ becomes

$$_2E_\phi = (j/4)\cos m\pi e^{-j\nu\phi} \int_{-\beta}^{\beta} \cosh mx$$

$$\exp\,[k\rho\,(\gamma^2 - 1)^{1/2}\cosh(x - j\psi)]\, e^{-2\nu f_x} dx \tag{7.38}$$

This contribution comes via K'_ν in (7.37), since it is replaced by the $H_\nu^{(2)\prime}$form in the range $-\beta < x < \beta$. It agrees with (7.25) when m is even but is of the wrong sign when m is odd. Hence we need to add a small additional term $_2\bar{E}_\phi$ for continuity; its value is found to be

$$_2\bar{E}_\phi = \tfrac{1}{2}j(1-\cos m\pi)e^{-j\nu\phi}\int_{-\beta}^{\beta} \cosh mx \exp\,[-k\rho(\gamma^2-1)^{1/2}$$

$$\sin\psi\,\sinh x]\ \pi kR_0\bar{\Gamma}_x J'_\nu[k(R_0 + \rho\cos\psi)\bar{\Gamma}_x]\, J_\nu(kR_0\bar{\Gamma}_x)\,dx \tag{7.39}$$

where

$$\bar{\Gamma}_x = [1-(\gamma^2 - 1)^{1/2}\sinh^2 x]^{1/2}$$

Note that it is always possible to add a term of this character since it is finite at the origin; hence a unique radiation condition at the torus cannot be found from a consideration of fields in $r < R_0$. But for $r > R_0$ the radiation condition at infinity uniquely gives (7.24) plus (7.25) for E_ϕ. The need for the extra term in (7.39) then follows in order to match at $\psi = \pi/2$. Its value is, of course, very small, since the asymptotic expansion of $J_\nu J'_\nu$ gives a factor $\exp(-2\nu f_x)$.

The calculations of this section do not affect the conclusions to be drawn from the radiation condition implied by (7.32) from a consideration of the range $r > R_0$ only. Rather, they indicate the nature of the field for $r < R_0$ necessary to give the required continuity of fields at the torus.

The nature of the sources to give these fields is discussed at length in an earlier paper[5] but will not be further considered here. (The reader interested in pursuing this facet should note that in Reference 5 a left-handed spherical coordinate system was inadvertently chosen. The error has been corrected in the present section but the corrected form has the effect of giving negative instead of positive z for $0 < \psi < \pi$. This should be borne in mind in any comparison of results.)

7.7 An alternative approach to the local radiation condition in cylindrical co-ordinates

As shown in section 7.3, the use of Debye's asymptotic expansion on the outgoing Hankel function leads directly to the appropriate form of the radiation condition. It was found that the total field in a local region near the curved wave-guiding surface consists of the unperturbed field of a flat surface and an exponential term whose amplitude decreases toward the curved surface. The additional term, although exceedingly small compared with the unperturbed field, nevertheless carries the information needed to calculate the attenuation caused by the curvature. The situation is of course more complicated for a toroidal structure. However, as evident from (7.32), the construction of a similar radiation condition is again a direct result of the Debye's expansion. In this section, we shall detail the nature of this expansion in order to gain some physical insight into the continuous radiation phenomenon of a curved structure.

We start by considering a suitable solution of a two-dimensional wave equation in cylindrical co-ordinates in the form of $\Phi(\rho,\phi)=\Phi(\rho)\exp(-j\nu\phi)$. The exponential term can be alternatively expressed as $\exp(-jk's)$ where $s=\phi R_0$ is the arc length and $k'=(\nu/R_0)$ is the propagation constant of a wave propagating along the curved structure. Accordingly $\Phi(\rho)$ satisfies the Bessel differential equation

$$\left[\frac{1}{\rho}\frac{\partial}{\partial\rho}\rho\frac{\partial}{\partial\rho}+k_0^2-(k'R_0/\rho)^2\right]\Phi(\rho) = 0 \qquad (7.40)$$

with solution $H_\nu^{(2)}(k_0\rho)$ satisfying the radiation condition at infinity. In (7.40), k_0 is the plane-wave propagation constant of the external space assumed here to be free-space. We now investigate the asymptotic behavior of this solution in regions away from the curved structure. For convenience, we also introduce a new co-ordinate system (u, s), given by $u=R_0\log(\rho/R_0)$ and $s=\phi R_0$. We note that for the region close to the curved surface, $(\rho-R_0)\ll R_0$, so that $u\simeq(\rho-R_0)$ is approximately a measure of radial distance away from the curved surface. Expressing the field in terms of u, we have from (7.40)

$$\left\{\frac{d^2}{du^2}+k_0^2[N^2(u)-\gamma^2]\right\}\Phi(u) = 0 \qquad (7.41)$$

where

$$N(u) = \exp(u/R_0), \qquad (7.42)$$

and $\gamma=k'/k_0$ is the normalized propagation constant. In the event $R_0\to\infty, N(u)=1$ and (7.41) gives precisely the transverse wave equation in rectangular co-ordinates. Evidently then, the effect of a finite curvature on waves propagating in a curved section, is the same as an inhomogeneous medium in a straight section, having a refractive index profile of $N(u)$. Now since the profile increases exponentially as exp (u/R_0), a propagating plane wave $(\gamma < 1)$ in the form of exp $\{-j k_0 [\gamma s + (N^2 - \gamma^2)^{1/2} u]\}$, which initially propagates at an angle $\theta = \cos^{-1}\gamma$ with respect to the s-axis, would tilt away from the s-axis and eventually propagate in a direction transverse to the s-axis. The situation is illustrated in Fig. 7.4, with arrows in 7.4*a* indicating the propagation direction of a wave in a curved region with a homogeneous medium, and in 7.4*b*, of a wave propagating in a straight section but with the inhomogeneous profile, $N(u)$.

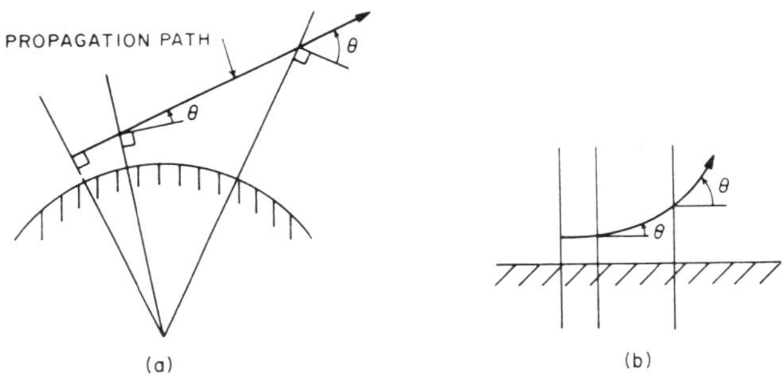

Fig. 7.4 Ray paths
a Plane wave on a curved section following a straight propagation path
b Equivalent ray on a straight inhomogeneous section following a curved propagation path

As we mentioned before, the application of interest here concerns the effect of curvature on surface waves along a dielectric structure. Typically, we assume $\gamma = k'/k_0 > 1$ and a solution of the form exp $(-[\gamma^2 - N^2]^{1/2} k_0 u)$ exp $(-jk_0 \gamma s)$ is adopted for the straight section. Thus, instead of a propagating plane wave, we have an evanescent field progressing in the radial direction. But because of the ever- increasing profile $N(u) = \exp(u/R_0)$, in the equivalent model, the rate of decay for the evanescent field has to decrease continuously as it moves away from the $u = 0$ surface. After it reaches the point $u_t = R_0 \log \gamma$ a different type of solution in the form of a propagating wave is needed since the relevant term, i.e. $(N^2 - \gamma^2)$ in (7.41), is now a positive number. As shown schematically, in Fig. 7.5, we therefore envisage an evanescent field progressing in the forward direction of

u and incident onto a boundary layer in the vicinity of the so-called turning-point or caustic at $u = u_t$ (or $\rho_t = \gamma R_0$ in the original cylindrical co-ordinates). It is then partially transmitted and transformed into a forward propagating wave, while at the same time partially reflected from the layer in the form of an evanescent wave progressing toward the dielectric structure. This then explains the existence of the additional term in the Debye's expansion of the outgoing Hankel function, $H_\nu^{(2)}(k_0\rho)$ as given in (7.6).

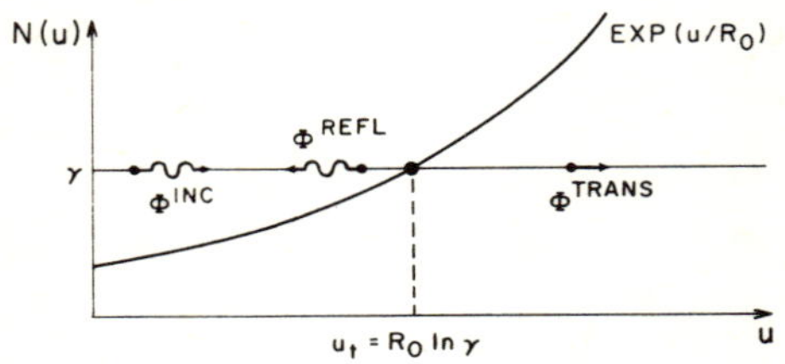

Fig. 7.5 Forward-progressing evanescent ray Φ^{inc} partially reflected as evanescent ray Φ^{refl} and partially transmitted as propagating ray Φ^{trans}

To determine more explicitly the amplitude of this reflected field in the local region near the curved surface $u = 0$, we first note that the refractive index $N(u)$ as given in (7.42) varies very slowly because of the large parameter R_0. Thus, in the region $u < u_t$, a WKB-type of approximate solution can be adopted;

$$\Phi(u) = A_0[(1-\gamma^2)/(N^2-\gamma^2)]^{1/4}\,\{e^{-k_0 Q(u)} + \mathcal{R}e^{k_0 Q(u)}\} \tag{7.43}$$

where

$$Q(u) = \int_0^u [-N^2(u)+\gamma^2]^{1/2}du \tag{7.44}$$

and $\mathcal{R}$ is a constant not yet determined. It is not difficult to see that the term $A_0 \exp(-k_0 Q)$ actually corresponds to an incident evanescent field progressing in the forward direction while $A_0 \mathcal{R} \exp(k_0 Q)$ is an evanescent field in the opposite direction. Now since the solution has to reduce to $\exp[-(\gamma^2-1)^{1/2}k_0 u]$ in the case when the curvature is infinite, we naturally expect the 'reflection coefficient' $\mathcal{R}$ to be a monotonically decreasing function of the radius of curvature R_0. Before deriving an expression for $\mathcal{R}$, we note that in the region beyond the turning-point $u > u_t$, a valid solution of (7.41) is

$$\Phi(u) = TA_0[(1-\gamma^2)/(N^2(u)-\gamma^2)]^{1/4}e^{-jk_0 P(u)} \tag{7.45}$$

where

$$P(u) = \int_{u_t}^{u} [N^2(u) - \gamma^2]^{1/2} du \tag{7.46}$$

with the 'transmission coefficient' T again undetermined. Here, however, we have the solution only in the form of a forward propagating wave, because of the outgoing wave requirement at infinity.

The integration in (7.44) can be performed analytically to give

$$Q(u) = -R_0 \left\{ \gamma \tanh^{-1} \frac{1}{\gamma} [\gamma^2 - N^2(u)]^{1/2} - [\gamma^2 - N^2(u)]^{1/2} \right\}_0^u \tag{7.47}$$

or, in cylindrical co-ordinates,

$$k_0 Q(u) = \nu [-f(k_0 \rho/\nu) + f(k_0 R_0/\nu)] \tag{7.48}$$

where $f(z)$ is defined in (7.4). For the region close to the curved surface, we can expand the function $f(k_0\rho/\nu)$ in a Taylor series about $\rho = R_0$ to yield

$$k_0 Q(u) = (\gamma^2 - 1)^{1/2} k_0 y; \qquad y = \rho - R_0 \tag{7.49}$$

so that the expression in (7.43) now takes a much simplified form:

$$\Phi(\rho) \simeq A_0 \{e^{-(\gamma^2-1)^{1/2} k_0 y} + \mathcal{R} e^{+(\gamma^2-1)^{1/2} k_0 y}\} \tag{7.50}$$

This confirms the earlier observation that the solution in this region should consist of an evanescent field progressing away from the curved surface $\rho = R_0$ and a reflected evanescent field progressing toward the surface.

To determine the explicit expression for $\mathcal{R}$, we need to recognize that the solution given in (7.43) for $u < u_t$ is connected to the solution given in (7.45) for $u > u_t$ through the region in the vicinity of the caustic. In this region, an approximate expansion for $Q(u)$ can be obtained by expanding it in a power series in a variable $\zeta^{1/2}$:

$$k_0 Q(u) = k_0 Q(u_t) - \frac{2}{3} \zeta^{3/2}; \quad \zeta = \left(\frac{k_0 R_0}{2\gamma^2} \right)^{2/3} [\gamma^2 - N^2(u)] \tag{7.51}$$

Consequently, a simplified version of (7.43) is then

$$\Phi(u) = \hat{A}_0 \zeta^{-1/4} [e^{\frac{2}{3}\zeta^{3/2}} + \hat{\mathcal{R}} e^{-\frac{2}{3}\zeta^{3/2}}]; \text{ for } \zeta > 0 \tag{7.52}$$

with

$$\hat{\mathcal{R}} = \mathcal{R} e^{2k_0 Q(u_t)}; \qquad k_0 Q(u_t) = \nu f(k_0 R_0/\nu), \tag{7.53}$$

and

$$\hat{A}_0 = A_0 \left(\frac{k_0 R_0}{2\gamma^2} \right)^{1/6} (1-\gamma^2)^{1/4} \, e^{-k_0 Q(u_t)}$$

Provided that the square root of ζ is defined as $+j|\zeta|^{1/2}$ when ζ is negative, we can also approximate (7.45) for the region $u > u_t$ in a similar manner

$$\Phi(u) = T\hat{A}_0 \zeta^{1/4} \, e^{-j\frac{2}{3}(-\zeta)^{3/2}}; \qquad \text{for } \zeta < 0. \qquad (7.54)$$

For most applications of interest, the parameter $k_0 R_0$ is usually extremely large. This allows us to assume that $|\zeta|$ is a large parameter even though the approximate solutions given in (7.52) and (7.54) are derived for the region near the caustic. In Appendix D, an asymptotic method for determining $\mathcal{R}$ is detailed and the following expression is obtained:

$$\mathcal{R} = -(j/2)e^{-2k_0 Q(u_t)} \qquad (7.55)$$

with

$$k_0 Q(u_t) = \nu f(k_0 R_0/\nu) = k_0 R_0 \gamma (\tanh^{-1} \Gamma - \Gamma)$$

The result derived here is identical to the previous one in (7.6) using explicitly the Debye's expansion.

Recalling the definition of $Q(u_t)$ in (7.44), it is clear that the exponential term in $\mathcal{R}$ corresponds exactly to twice the total attenuation of an evanescent field from the curved surface at $u = 0$ to the location of the caustic. This then suggests the possibility of minimizing the curvature effect simply by relocating the turning-point. As is evident from (7.41) and (7.42), a complete removal of this point is possible, at least in principle, whenever the surrounding medium has a refractive index profile of $\exp(-u/R_0)$, or, in the cylindrical co-ordinates, a profile of R_0/ρ. Since the combined equivalent profile $N(u)$ is now held at unity for all u, this means a profile inversely proportional to ρ would ideally produce no radiation. Even though such a profile is not a physically plausible one because of the vanishingly small refractive index at infinity, our discussion here does point out how one could achieve a reduction, if not a complete removal, of the curvature effect.

A case of interest in optical waveguide application is when the normalized propagation constant of the surface-wave is very close to unity, i.e. $|\gamma - 1| \ll 1$. Since the term $(\tanh^{-1}\Gamma - \Gamma)$ can be approximated by $\Gamma^3/3$ in this case, we have from (7.55) a simplified expression for $\mathcal{R}$ (provided that $k_0 R_0 \Gamma^5/5 \ll 1$)

$$\begin{aligned} \mathcal{R} &= (-j/2) \exp\left[-\tfrac{2}{3} k_0 R_0 \gamma^{-2} (\gamma^2 - 1)^{3/2}\right] \\ &= (-j/2) \exp\left[-2(\gamma^2 - 1)^{1/2} k_0 D_e\right] \qquad (7.56) \end{aligned}$$

where $D_e \simeq 2R_0(\gamma - 1)/3 = 2(\rho_t - R_0)/3$ and $\rho_t = R_0\gamma$ is the location of the caustic. Thus, D_e is approximately two thirds of the total distance measured from the curved surface to the caustic. Substitution of (7.56) into (7.50) now yields a surprisingly simple form for the field component near the curved surface

$$\Phi(y) = A_0\{e^{-(\gamma^2-1)^{1/2}k_0 y} - (j/2)e^{-(\gamma^2-1)^{1/2}k_0(2D_e-y)}\} \quad (7.57)$$

with $y = \rho - R_0$ being the distance measured from the curved surface. Equation (7.57) lends itself to a very interesting physical interpretation. In addition to the self-field of a straight guide, the field component near the curved surface appears to have an additional contribution due to an image having an amplitude equal to $(-j/2)$ of the original one, and located at a distance not quite twice the distance between the original structure and the caustic. The fact that the separation between the two structures is not exactly twice the distance to the caustic can be traced directly to the non-uniform nature of the decay of the evanescent field in an inhomogeneous medium; that the amplitude of the image differs from unity is due to the specific reflection mechanism of the caustic. From this viewpoint, we can therefore consider that the effect of the curvature is to produce an image structure running parallel to the original one. It is then not difficult to understand why the local field behavior of a straight guide needs to be modified whenever a finite curvature is introduced.

7.8 Local radiation condition for a curved structure of arbitrary but finite cross-section

We now proceed to discuss the local radiation condition of a curved structure having arbitrary but finite cross-section. The structure is located in a global cylindrical co-ordinate system (r, ϕ, z) as shown in Fig. 7.6. For the present, the only restriction we impose is that in the exterior region $r > r_{max}$ (where r_{max} is the largest value of r in the guide cross-section) the medium be homogeneous with a refractive index n_0, different from 1 for generality. In this region, a Cartesian field component Φ must satisfy the scalar wave equation:

$$\left[\frac{1}{r}\frac{\partial}{\partial r}r\frac{\partial}{\partial r} + \frac{\partial^2}{\partial z^2} + k^2 - (k'R_0/r)^2\right]\Phi(r, z) = 0 \quad (7.58)$$

and once again an $\exp(-j\nu\phi)$ or alternatively $\exp(-jk's)$ form of propagation along the curved structure, with arc length $s = R_0\phi$, and $k' = \nu/R_0$, $k = k_0 n_0$, has been assumed. Taking advantage of the uniformity of the region, we may define a Fourier transform pair:

$$\Phi(r,z) = \int_{-\infty}^{\infty} \tilde{\Phi}(r;\xi)\, e^{-jk\xi z} d\xi$$

$$\tilde{\Phi}(r;\xi) = \frac{k}{2\pi}\int_{-\infty}^{\infty} \Phi(r,z) e^{jk\xi z}\, dz \qquad (7.59)$$

so that the spectrum function $\tilde{\Phi}$ satisfies the ordinary differential equation

$$\left[\frac{1}{r}\frac{\partial}{\partial r} r \frac{\partial}{\partial r} + k^2(1-\xi^2) - (k'R_0/r)^2\right] \tilde{\Phi}(r;\xi) = 0 \qquad (7.60)$$

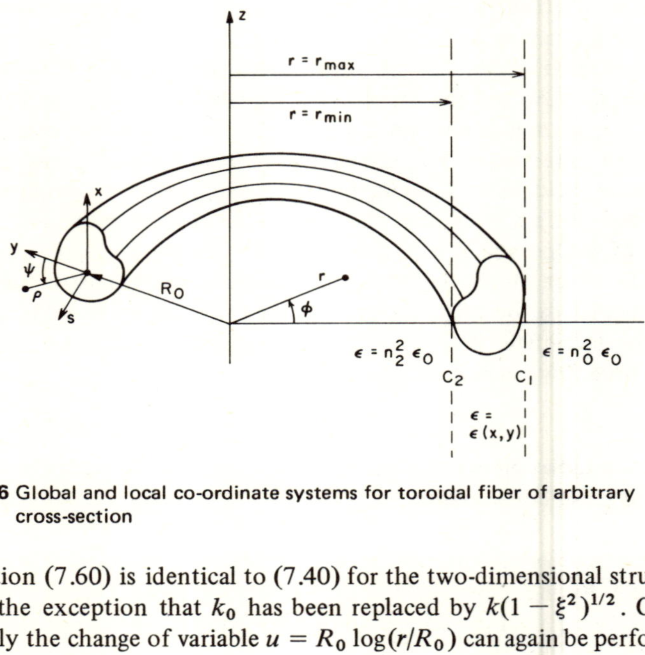

Fig. 7.6 Global and local co-ordinate systems for toroidal fiber of arbitrary cross-section

Equation (7.60) is identical to (7.40) for the two-dimensional structure with the exception that k_0 has been replaced by $k(1-\xi^2)^{1/2}$. Consequently the change of variable $u = R_0 \log(r/R_0)$ can again be performed to yield

$$\left\{\frac{d^2}{du^2} + k_0^2\,[n_0^2 \tilde{N}^2(u;\xi) - \bar{\gamma}^2]\right\} \tilde{\Phi}(u;\xi) = 0 \qquad (7.61)$$

where $\bar{\gamma} = k'/k_0 = n_0\gamma$

so that an equivalent refractive index profile $n_0\tilde{N}(u;\xi)$ appears:

$$\tilde{N}(u;\xi) = (1-\xi^2)^{1/2} e^{u/R_0}$$

When $R_0 \to \infty$, the solution of (7.61) must reduce to

$$\tilde{\Phi}_0(u;\xi) = \tilde{A}_0(\xi) e^{-\eta(\bar{\gamma},\xi)k_0 y} \qquad (7.62)$$

where

$$\eta(\bar{\gamma}, \xi) = [\bar{\gamma}^2 + n_0^2(\xi^2 - 1)]^{1/2} \tag{7.63}$$

$y = r - R_0$ represents the radial distance measured from the curved surface, and $\tilde{A}_0$ is a spectral amplitude. Thus, before curvature is introduced, the spectrum representation of Φ for a straight guiding structure is

$$\Phi_0(y, z) = \int_{-\infty}^{\infty} \tilde{A}_0(\xi) e^{-\eta(\bar{\gamma},\xi)k_0 y} e^{-jk_0 n_0 \xi z} \, d\xi \tag{7.64}$$

(we have used y here instead of r as the argument of Φ_0 since both r and R_0 have become infinite while y remains finite for the straight guide).

Now in order to take into account properly the finite curvature it is important to realize that each spectral component has to be treated independently since the equivalent profile changes according to the value that ξ takes. In the range where $|\xi| > 1$, the term $n_0^2\tilde{N}^2 - \bar{\gamma}^2$ in (7.61) becomes increasingly negative so as to yield a more and more rapidly decaying field as it moves away from the curved surface. Therefore, no caustic is encountered and no reflection is possible in this case. On the other hand, in the range where $|\xi| \leqslant 1$, $\tilde{N}^2$ is positive near $y = 0$ and a caustic always exists, even though its location moves away to infinity as the magnitude of ξ approaches unity. Following the discussion in the preceding section, we can write down the expression for the total field in the region near the curved structure as

$$\Phi(r, z) = \Phi_0(y, z) + \int_{-1}^{1} \tilde{A}_0(\xi)\tilde{\mathcal{R}}(\xi) e^{\eta(\bar{\gamma},\xi)k_0 y} \, d\xi \tag{7.65}$$

with the 'reflection-coefficient' $\tilde{\mathcal{R}}(\xi)$ given according to (7.55) as

$$\tilde{\mathcal{R}}(\xi) = (-j/2) e^{-2k_0 \tilde{Q}(u_t;\xi)}$$

$$\tilde{Q}(u_t; \xi) = R_0\{\bar{\gamma} \tanh^{-1} [\eta(\bar{\gamma}, \xi)/\bar{\gamma}] - \eta(\bar{\gamma}, \xi)\} \tag{7.66}$$

Obviously, (7.65) in form is the same for $n_0 = 1$ as the expression for $E_\phi = E_1 + E_2$ given by (7.24) and (7.25) in section 7.5 where specifically the construction of the local radiation condition for a curved cylindrical fiber was discussed. The emphasis here is that (7.65) is actually applicable to a curved structure of arbitrary cross-section. It points out that the leading term in the total field expression near the structure is always that of the straight structure. Thus, provided that we know the spectrum representation of this field component, the local radiation condition can always be properly implemented by including the reflected evanescent field from caustics in that part of the spectrum between $\xi = \pm 1$. Now with the realization that the reflection coefficient

$\tilde{\mathfrak{R}}$ usually involves a large parameter k_0R_0 in its exponent, one can further evaluate the integral in (7.65) using a standard steepest-descent method. It is easy to verify that the saddle-point defined by $\partial\tilde{Q}(u_t;\xi)/\partial\xi = 0$ is located at $\xi = 0$, and that the steepest-descent path is essentially on the real axis between $\xi = \pm 1$ with the exponent going to $-\infty$ at $\xi = \pm 1$. Thus, by expanding $\tilde{Q}$ in a Taylor series around the saddle-point, we have from (7.65) the relation

$$\Phi(r,z) \sim \Phi_0(y,z) - (j/2)\tilde{A}_0(0)e^{\gamma\Gamma ky}\, e^{-2kR_0\gamma(\tanh^{-1}\Gamma - \Gamma)} \cdot \int_{-\infty}^{\infty} e^{-kR_0\gamma\Gamma u^2}\, du$$

where $\Gamma = (1-\gamma^{-2})^{1/2}$ and $\gamma = \bar{\gamma}/n_0\,; k = n_0k_0$.
Hence

$$\Phi(r,z) \sim \Phi_0(y,z) + \sigma_0\left(\frac{\pi}{\gamma\Gamma kR_0}\right)^{1/2} \tilde{A}_0(0)e^{\gamma\Gamma ky} \tag{7.67}$$

$$\sigma_0 = \tilde{\mathfrak{R}}(0) = -(j/2)e^{-2\tau}; \qquad \tau = kR_0(\tanh^{-1}\Gamma - \Gamma) \tag{7.68}$$

Next we discuss the physical significance of this result derived for a curved structure of arbitrary cross-section. It is clear from (7.67) that the field component in the local region near the curved structure consists of a primary field corresponding to the unperturbed case of a straight guiding structure, and a secondary field having the appearance of a reflected evanescent field progressing in the negative y direction. The expression $\tilde{A}_0(0)$ which appears in the amplitude of the reflected field is in fact related to the forward pattern of the incident field. Specifically, one can start from the spectrum representation in (7.64) and employ a standard steepest-descent method to show that the far-field expression of $\Phi_0(y,z)$ is given as

$$\Phi_0(y,z) \simeq \left[\frac{2\pi(\gamma^2-1)^{1/2}}{k\rho}\right]^{1/2} F(\psi)e^{-k\rho(\gamma^2-1)^{1/2}} \tag{7.69}$$

$$F(\psi) = \cos\psi\,\tilde{A}_0(\xi)\,|_{\xi=-j(\gamma^2-1)^{1/2}\sin\psi}$$

Thus, the amplitude of the reflected field is directly proportional to the strength of the primary field in the forward radial direction, i.e. $\tilde{A}_0(0) = F(\psi)$ for $\psi = 0$. Moreover, the expression for σ_0 contains the factor $\exp(-2kR_0\gamma\,[\tanh^{-1}\Gamma - \Gamma])$, which, as evident from (7.44) and (7.55), is precisely the decay of the primary field in the forward direction over a distance equal to twice the separation between the curved surface and the location of the caustic in the same direction.

As mentioned before, a case of interest in the fiber application is when the normalized propagation constant γ is near unity, i.e. $|\gamma - 1|$

$\ll 1$. For a two dimensional structure, one recalls that the effect of curvature is to introduce an image structure located at a distance of $2D_e$ away, which, from the definition of D_e in (7.56), is somewhat smaller than twice the separation between the curved surface and the caustic. Now, assuming the image structure has a far-field expression identical with the real one, the substitution of $\rho = 2D_e \simeq 2R_0(\gamma^2 - 1)/3$ into (7.69) yields

$$\Phi_0(2D_e, 0) \simeq \left[\frac{3\pi}{k_0 R_0(\gamma^2 - 1)^{1/2}}\right]^{1/2} \tilde{A}_0(0) e^{-2k_0 R_0(\gamma^2 - 1)^{3/2}/3}$$

Thus, in the case of a curved fiber of finite cross-section, the total field component near the fiber as given in (7.67) and (7.68) can be further simplified to

$$\Phi(r,z) = \Phi_0(y,z) + (-j/2\sqrt{3})\Phi_0(2D_e - y, 0). \qquad (7.70)$$

Equation (7.70) can then be interpreted in the same manner as (7.57) in the two-dimensional case, with the only exception that the strength of the image field is now $(-j/2\sqrt{3})$ relative to the original guide.

Up until now, our discussion on the radiation condition in the local region near a curved structure has been restricted mainly to one of the Cartesian components of electromagnetic field, say $E_z(r, z)$ or $H_z(r, z)$. In practice it is usually not very convenient to use these components in a problem characterized by other coordinate systems. Extension to other field components in a cylindrical coordinate system in principle can be achieved by relating the spectrum representation of these components in terms of $\tilde{E}_z$ and $\tilde{H}_z$ by

$$\begin{aligned}
\tilde{E}_\phi(r;\xi) &= \frac{1}{n_0^2(1-\xi^2)}\left[\frac{j\zeta_0}{k_0}\frac{\partial}{\partial r}\tilde{H}_z(r;\xi) - \frac{n_0 \xi R_0 \bar{\gamma}}{r}\tilde{E}_z(r;\xi)\right] \\
\tilde{H}_\phi(r;\xi) &= \frac{1}{n_0^2(1-\xi^2)}\left[\left(\frac{-jn_0^2}{k_0\zeta_0}\right)\frac{\partial}{\partial r}\tilde{E}_z(r;\xi) - \frac{n_0 \xi R_0 \bar{\gamma}}{r}\tilde{H}_z(r;\xi)\right] \\
\tilde{E}_r(r;\xi) &= \frac{1}{n_0^2(1-\xi^2)}\left[\left(\frac{-jn_0\xi}{k_0}\right)\frac{\partial}{\partial r}\tilde{E}_z(r;\xi) - \zeta_0\frac{R_0\bar{\gamma}}{r}\tilde{H}_z(r;\xi)\right] \\
\tilde{H}_r(r;\xi) &= \frac{1}{n_0^2(1-\xi^2)}\left[\left(\frac{-jn_0\xi}{k_0}\right)\frac{\partial}{\partial r}\tilde{H}_z(r;\xi) - \frac{n_0^2}{\zeta_0}\left(\frac{R_0\bar{\gamma}}{r}\right)\tilde{E}_z(r;\xi)\right]
\end{aligned} \qquad (7.71)$$

(Here, $\zeta_0 = (u_0/\epsilon_0)^{1/2}$ is the intrinsic impedance of a plane wave in free-space.) Now since the approach we used to find $\tilde{\Phi}$ applies to both $\tilde{E}_z$ and $\tilde{H}_z$, we have immediately from the discussion leading to (7.65) the following expression

$$\begin{Bmatrix} \tilde{E}_z(r;\xi) \\ \tilde{H}_z(r;\xi) \end{Bmatrix} = \begin{Bmatrix} \tilde{E}_{z0}(\xi) \\ \tilde{H}_{z0}(\xi) \end{Bmatrix} [e^{-\eta(\bar{\gamma},\xi)k_0 y} + \tilde{\mathcal{R}}(\xi)\, U(\xi)\, e^{+\eta(\bar{\gamma},\xi)k_0 y}] \tag{7.72}$$

where $\tilde{E}_{z0}$, $\tilde{H}_{z0}$ are respectively the amplitude of the spectrum functions for E_{z0} and H_{z0} of a straight guide; $U(\xi)$ is a step function which equals unity for $|\xi| < 1$ and zero, otherwise, and $\tilde{\mathcal{R}}(\xi)$ is the reflection coefficient defined in (7.66). Substitution of (7.72) into (7.71), followed by a steepest-descent evaluation of the inverse Fourier transform in (7.59) for each field component finally yields

$$\begin{Bmatrix} E(r,z) \\ H(r,z) \end{Bmatrix} = \begin{Bmatrix} E_0(y,z) \\ H_0(y,z) \end{Bmatrix} + \begin{Bmatrix} \delta E(y,0) \\ \delta H(y,0) \end{Bmatrix}, \tag{7.73}$$

for $y = r - R_0$, where the perturbations to the two longitudinal components, i.e. δE_z and δH_z are obtained directly from (7.67) in terms of the spectrum amplitude $\tilde{E}_{z0}(0)$ and $\tilde{H}_{z0}(0)$ as

$$\begin{Bmatrix} \delta E_z(y,0) \\ \delta H_z(y,0) \end{Bmatrix} = -(j/2)\left(\frac{\pi}{\gamma\Gamma k R_0}\right)^{1/2} \begin{Bmatrix} \tilde{E}_{z0}(0) \\ \tilde{H}_{z0}(0) \end{Bmatrix} e^{-2T} e^{(\gamma^2-1)^{1/2} ky} \tag{7.74}$$

The perturbations to the other components are then derived from (7.71) to be

$$\left.\begin{aligned} \delta E_\phi(y,0) &= \left(\frac{j\zeta_0}{n_0^2 k_0}\right)\frac{\partial}{\partial y}[\delta H_z(y,0)] \\ \delta H_\phi(y,0) &= \left(\frac{-j}{k_0\zeta_0}\right)\frac{\partial}{\partial y}[\delta E_z(y,0)] \\ \delta E_r(y,0) &= \left(\frac{-\zeta_0 R_0\bar{\gamma}}{n_0^2[R_0+y]}\right)[\delta H_z(y,0)] \\ \delta H_r(y,0) &= \left(\frac{-R_0\bar{\gamma}}{\zeta_0[R_0+y]}\right)[\delta E_z(y,0)] \end{aligned}\right\} \tag{7.75}$$

Thus, the local radiation condition as applied to the field components in any given coordinate system can now be obtained from (7.73) – (7.75). In particular, when the curved structure is locally characterized by a toroidal system to represent the bending of a round fiber, the result given here obviously has to reduce to the one previously constructed in (7.32). Before we can demonstrate the equivalence of these two results, we first need to find the spectral representation of $K_m\,[k\rho(\gamma^2-1)^{1/2}]\exp(-jm\psi)$, which forms the basic solution of the scalar wave equation for a straight cylindrical structure that satisfies the necessary radiation condition at infinity. This is done in Appendix E and the result is as follows:

$$K_m\,[k\rho(\gamma^2-1)^{1/2}]\,\exp(-jm\psi)$$

$$=\int_{-\infty}^{\infty}\frac{1}{2(\gamma^2+\xi^2-1)^{1/2}}\left[\frac{\xi+(\gamma^2+\xi^2-1)^{1/2}}{(\gamma^2-1)^{1/2}}\right]^m e^{-(\gamma^2+\xi^2-1)^{1/2}ky}\,e^{-jk\xi z}\,d\xi \tag{7.76}$$

where again $\rho=(y^2+z^2)^{1/2}$ and $\psi=\tan^{-1}(-z/y)$. Now if we assume the field component $E_{\phi o}(y,z)$ is given in the form of (7.76) when the structure is straight (i.e. both r and R_0 approach infinity but $y=r-R_0$ remains finite), the spectral amplitude of $H_{zo}(y,z)$ at $\xi=0$ is then related to that of $E_{\phi o}$ by (7.71) and is simply

$$\widetilde{H}_{zo}(y,0)=\frac{jn_0}{2\zeta_0(\gamma^2-1)}\widetilde{E}_{\phi o}(y,0) \tag{7.77}$$

A substitution of (7.77) into (7.74) using (7.75), yields an explicit expression for δE_ϕ as

$$\delta E_\phi(y,0)=(j/4)\left(\frac{\pi}{kR_0}\right)^{1/2}(\gamma^2-1)^{-3/4}e^{-2\tau}e^{(\gamma^2-1)^{1/2}ky} \tag{7.78}$$

Hence, with the use of the definition in (7.73) and the well-known identity

$$\exp(k\gamma\Gamma y)=\sum_{m=0}^{\infty}\epsilon_m I_m(k\gamma\Gamma\rho)\cos m\psi$$

where $\epsilon_0=1$ and $\epsilon_m=2$ for $m>0$, we finally obtain

$$E_\phi(r,z)=K_m(k\gamma\Gamma\rho)\exp(-jm\psi)+\sigma\sum_{m=0}^{\infty}\epsilon_m I_m(k\gamma\Gamma\rho)\cos m\psi \tag{7.79}$$

where σ has the same value as in (7.31). A further splitting of (7.79) into symmetric and antisymmetric parts will then produce the identical expressions previously constructed in (7.32) for the two independent solutions in a toroidal system.

References

1 Chang, D. C. and Barnes, F. S.: 'Reduction of radiation loss in a curved dielectric slab waveguide'. Report No. 2, Electromagnetics Laboratory, Department of Electrical Engineering, University of Colorado, Boulder, Colorado, 1973
2 Marcuse, D.: 'Light transmission optics' (Van Nostrand-Reinhold, New York, 1972), pp. 400–404
3 Snyder, A. W., White, I., and Mitchell, D. J.: 'Radiation from bent optical waveguides', *Electronics Letters,* Vol. 11, No. 15, July 1975, pp. 332–333
4 Abramowitz, M., and Stegun, I. A., (ed.): 'Handbook of mathematical functions' (Applied Mathematics Series 55) Washington, D.C., NBS, June 1964, pp. 366 and 378
5 Lewin, L.: 'Radiation from curved dielectric slabs and fibers', *IEEE Trans. MTT*, Vol. 22, No. 7, July 1974, pp. 718–727

Chapter 8

The curved dielectric slab

8.1 Solution in cylindrical co-ordinates

A slab of width b, relative permittivity ϵ_r and radius of curvature R_0 is shown in Fig. 8.1. Three regions can be distinguished. For $\rho < R_1$, the

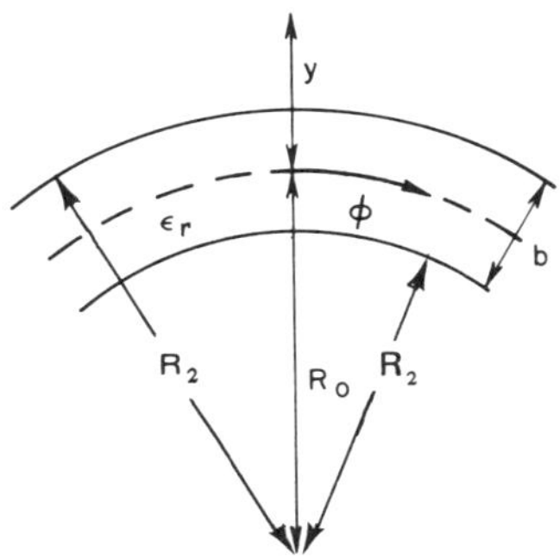

Fig. 8.1 Curved dielectric slab

free space below the concave surface of the slab, solutions for the component H_z of the magnetic field, for the case of a TM mode, can be obtained from $e^{-j\nu\phi}J_\nu(k_0\rho)$, since the field must be finite at $\rho = 0$. In the interior of the slab $R_1 < \rho < R_2$ terms in both $J_\nu(k\rho)$ and $Y_\nu(k\rho)$ with $k = k_0\epsilon_r^{1/2}$, are needed. And for $\rho > R_2$ outgoing waves in free space are needed, varying as $H_\nu^{(2)}(k_0\rho)$. If A, B, C and D are arbitrary multipliers, then the axial magnetic field in the three regions can be written

$$H_z = \begin{cases} e^{-j\nu\phi}AJ_\nu(k_0\rho) & \rho < R_1 \\ e^{-j\nu\phi}[BJ_\nu(k\rho) + CY_\nu(k\rho)] & R_1 < \rho < R_2 \\ e^{-j\nu\phi}DH_\nu^{(2)}(k_0\rho) & \rho > R_2 \end{cases} \quad (8.1)$$

Continuity at $\rho = R_1$ and R_2 leads to the two relations

$$\begin{aligned} AJ_\nu(k_0R_1) &= BJ_\nu(kR_1) + CY_\nu(kR_1) \\ DH_\nu^{(2)}(k_0R_2) &= BJ_\nu(kR_2) + CY_\nu(kR_2) \end{aligned} \tag{8.2}$$

The E_ϕ component of electric field is proportional to $(1/\epsilon)\partial H_z/\partial\rho$ and must also be continuous at $\rho = R_1$ and R_2. These conditions are sufficient to ensure the required continuity of the remaining field quantities, and lead to

$$\begin{aligned} \epsilon_r^{1/2} AJ'_\nu(k_0R_1) &= BJ'_\nu(kR_1) + CY'_\nu(kR_1) \\ \epsilon_r^{1/2} DH_\nu^{(2)\prime}(k_0R_2) &= BJ'_\nu(kR_2) + CY'_\nu(kR_2) \end{aligned} \tag{8.3}$$

Eliminating A, B, C and D from these equations gives

$$\frac{J_\nu(k_0R_1)J'_\nu(kR_1) - \epsilon_r^{1/2}J_\nu(kR_1)J'_\nu(k_0R_1)}{J_\nu(k_0R_1)Y'_\nu(kR_1) - \epsilon_r^{1/2}Y_\nu(kR_1)J'_\nu(k_0R_1)} = \frac{H_\nu^{(2)}(k_0R_2)J'_\nu(kR_2) - \epsilon_r^{1/2}J_\nu(kR_2)H_\nu^{(2)\prime}(k_0R_2)}{H_\nu^{(2)}(k_0R_2)Y'_\nu(kR_2) - \epsilon_r^{1/2}Y_\nu(kR_2)H_\nu^{(2)\prime}(k_0R_2)} \tag{8.4}$$

This is an equation for the determination of ν, and hence the propagation and attenuation along the slab. The needed solutions are for ν large and complex. There is no known analytic solution to (8.4), though Marcatili[1] used asymptotic expansions to obtain an approximate solution for small curvatures. Dang[2] has produced a computer program for solving (8.4) numerically, based on a calculation of the values of Bessel functions through a numerical evaluation of Hankel's integrals. Since ν will be large but may involve only a very small imaginary part, very high accuracy is necessary in the computations to calculate the attenuation accurately. The results can be compared to values extracted from perturbation solutions but differences may be difficult to interpret due to the unknown accuracy of the numerical computations. Some comparisons of this nature are made in section 8.3.

8.2 Perturbation solution

Putting $\chi = 1/R_0$, the curvature at the slab center, a perturbation solution in the curvature can be written as

$$H_z = e^{-jk's}H \tag{8.5}$$

where

$$H = H_0 + \chi H_1 + \chi^2 H_2 + \ldots \tag{8.6}$$

$$k' = k_0'(1 + \chi b B_1 + \ldots) \tag{8.7}$$

Here it is rather implied by the notation that terms like H_0, k_0', B_1 etc., are independent of χ and reduce to the values for the straight slab as $\chi \to 0$. However the analysis, through the use of the radiation condition (7.6), also brings in exponential terms. These decay to zero faster than any power of χ, and the only justification for retaining them is for their quadrature component, which ultimately gives rise to the radiation. It is this component, as will be shown, that is contained implicitly in the various coefficients in the above equations. We also define here the following quantities:

$$\begin{aligned} \gamma^2 &= k_0^2\epsilon_r - k_0'^2 = k^2 - k_0'^2 \\ \delta^2 &= k_0'^2 - k_0^2 \\ \psi &= b\gamma/2 \end{aligned} \tag{8.8}$$

From (4.1) the Helmholtz equation in local co-ordinate form is

$$(1+\chi y)^2 \partial^2 H/\partial y^2 + \chi(1+\chi y)\partial H/\partial y + [(1+\chi y)^2 k_0^2(\epsilon_r, 1) - k'^2]H = 0 \tag{8.9}$$

where ϵ_r or 1 is taken according as to whether the field is being considered inside or outside the slab.

For the zero-order solution we get the equation

$$\partial^2 H_0/\partial y^2 + [k_0^2(\epsilon_r, 1) - k_0'^2]H_0 = 0 \tag{8.10}$$

Noting that for $y < -b/2$, a simple decaying term is needed whilst, for $y > b/2$, (7.6) requires also an increasing term in a certain ratio at the slab upper surface, we get the forms

$$H_0 = \begin{cases} Ae^{\delta(y+b/2)}, & y < -b/2 \\ B\cos\gamma y + C\sin\gamma y, & -b/2 < y < b/2 \\ D\{e^{-\delta(y-b/2)} + Ke^{\delta(y-b/2)}\}, & y > b/2 \end{cases} \tag{8.11}$$

Here, A, B, C and D are constants to the determined, whilst K can be found from (7.6) and is given by

$$K = -\tfrac{1}{2}j\exp\left\{\delta b - 2R_0\left[k_0' \log\frac{\delta + k_0'}{k_0} - \delta\right] - 2B_1 k_0' b \log\frac{\delta + k_0'}{k_0}\right\} \tag{8.12}$$

The term δb in (8.12) appears because y in this section is measured from the slab center-line, whilst in (7.6) the ratio of the two terms is determined at the upper surface from which y in that section was measured. The displacement gives the δb term. B_1 also appears in (8.12) because $\nu = k_0'(R_0 + bB_1)$ to first order, so a contribution of $0(1)$ appears if B_1 itself has a significant magnitude.

Field matching requires H_0 continuous at $y = \pm b/2$, and since E_ϕ, proportional to $(1/\epsilon)\partial H_z/\partial y$, is also continuous there we have four relations between A, B, C and D. On eliminating these amplitudes we get the secular equation

$$\tan^2\psi + \tan\psi\,[(K-1)\Delta + (K+1)/\Delta] - 1 = 0 \qquad (8.13)$$

where $\psi = b\gamma/2$ and $\Delta = \epsilon_r\delta/\gamma$. This equation determines a further relation between the constants of (8.8) from which k' can be ultimately determined.

To solve (8.13) we can write in the form

$$[\tan\psi - \Delta(1-K)]\,[\tan\psi + (1+K)/\Delta] = K^2 \qquad (8.14)$$

Now for large R_0, K is exponentially small, so K^2 in (8.14) can be neglected, giving

$$\tan\psi = \Delta(1-K) \quad \text{or} \quad (1+K)/\Delta \qquad (8.15)$$

The first relation corresponds to the symmetrical mode in the straight slab when $K = 0$, and will be pursued further here. If k_{00}' is the solution corresponding to $K = 0$, then an expansion around this value gives

$$k_0' = k_{00}' + KL \qquad (8.16)$$

where

$$L = \frac{\epsilon_r}{\epsilon_r - 1} \cdot \frac{2\gamma_0^2\delta_0^2}{(2\epsilon_r k_0^2 + \delta_0 b k_{00}'^2)k_{00}'} \qquad (8.17)$$

and γ_0 and δ_0 come from writing k_{00}' for k_0' in (8.8). In this relation k_{00}' is the solution obtained from the secular equation

$$\tan(b\gamma_0/2) = \epsilon_r\delta_0/\gamma_0 \qquad (8.18)$$

In calculating K it is sufficient, to this order of accuracy, to use δ_0 and γ_0 in (8.12).

8.3 First-order solution

In order to complete the above solution we need the value of B_1 for insertion into (8.12). As it happens, B_1 turns out to be exponentially small so that (8.16) and (8.17) do give k_0' accurately without more ado. However, this assertion is far from obvious, and in any case B_1 is needed in (8.7) for k'. From a consideration of the coefficient of χ in (8.6) we get

$$\partial^2 H_1/\partial y^2 + [k_0^2(\epsilon_r, 1) - k_0'^2]H_1 = -\partial H_0/\partial y + k_0'^2 H_0(B_1 - 2y) \tag{8.19}$$

If the amplitude B in (8.11) is arbitrarily taken as unity, then the solution of (8.19) for H_1 takes the form

$$H_1 = \begin{cases} e^{\delta(y+b/2)}\cos\psi[M + y(B_1 + k_0^2/\delta k_0'^2) - y^2]k_0'^2/2\delta, \\ y < -b/2 \\ -(k_0^2\epsilon_r/2\gamma^2)y\cos\gamma y + (k_0'^2/2\gamma)(N + B_1 y - y^2)\sin\gamma y, \\ -b/2 < y < b/2 \\ (\cos\psi k_0'^2/2\delta)\{e^{-\delta(y-b/2)}[y^2 - y(B_1 - k_0^2/\delta k_0'^2)] + P[e^{-\delta(y-b/2)} \\ + Ke^{\delta(y-b/2)}]\}, \quad y > b/2 \end{cases} \tag{8.20}$$

with B_1, M, N and P to be determined by matching H_z and E_ϕ at $y = \pm b/2$. Use has been made of the radiation condition in local form in determining the form of the terms in M and P.

The calculation of B_1 in this way is lengthy but straightforward, and gives $B_1 = K\beta$ where

$$\beta = \frac{\epsilon_r\delta^2[(1+X)(\epsilon_r^2 - X\gamma^2/\delta^2) + \epsilon_r^3\delta^2/k_0'^2]/4X}{X[(\epsilon_r - 1)(\epsilon_r k_0'^2 - \gamma^2) + K\{\gamma^2 - \epsilon_r(\epsilon_r - 1)k_0'^2\}] + \epsilon_r[k_0'^2(\epsilon_r - 1) + K\gamma^2]} \tag{8.21}$$

and

$$X = \delta b(\epsilon_r\delta^2 + k_0'^2)/2k_0^2 \tag{8.22}$$

To the extent that K is negligible, the denominator in (8.21) simplifies to $(\epsilon_r - 1)[\epsilon_r k_0'^2 + X(\epsilon_r k_0'^2 - \gamma^2)]$.

Thus B_1 contains K as a factor and since K decreases exponentially with R_0, $B_1 \to 0$ as $R_0 \to \infty$, and its effect on (8.12) is usually negligible. Since $\nu = k'R_0 = k_0'(R_0 + bB_1)$ to first order in $1/R_0$, we get, to first order in K and $1/R_0$,

$$k' = k_{00}' + K(L + \beta k_{00}' b/R_0) \tag{8.23}$$

The quantities L and β are given in (8.17) and (8.21). To take the approximation further would be extremely tedious, but (8.23) should be accurate even for quite modest values of R_0. Thus, with $\lambda_0 = 1$ cm, $b = 0{\cdot}4$ mm, $\epsilon_r = 4$ and $R_0 = 5b$, the value of ν is found as $21{\cdot}06 - j0{\cdot}0001$. The corresponding value from Dang's computation[2] is $20{\cdot}89 - j0{\cdot}00004$. The real parts are quite close but the imaginary parts, which give the attenuation, are too small to compare accurately. A closer agreement is obtained with the extreme case of a very leaky second-order mode with $b = 0{\cdot}8\,\text{mm}$ and $R_0 = 2b$. The two values are, respectively, $11{\cdot}628 - j5{\cdot}064$ and $12{\cdot}410 - j5{\cdot}139$. For comparison, the first-order mode in this case has the imaginary parts of ν calculated as $8{\cdot}10^{-5}$ and $5{\cdot}10^{-6}$ respectively.

8.4 Calculation of attenuation constant based on a power balance concept

As is evident from the perturbation solution presented in (8.12), (8.16) and (8.17), the essential difference between the propagation constant of a surface wave propagating along a curved slab and that on a straight slab is a negative, but purely imaginary number equal to $k_0' - k_{00}' = KL$. Thus, while the field distribution as well as phase velocity remain largely unchanged from a straight slab, attenuation of the surface wave now occurs as a result of the finite curvature. Recalling the earlier discussion on the local radiation condition in Chapter 7, it can clearly be seen that the attenuation of the surface wave is a direct consequence of the continuous radiation beyond the caustic when the structure is curved. In fact, the amount of power loss per unit arc length can be calculated from the time-average Poynting vector, in the radial direction as

$$P_{rad} = \tfrac{1}{2}\,\mathrm{Re}\,(E_\phi H_z^*)$$

which according to (8.11) and the expression $E_\phi = (j/\omega\epsilon_0)\partial H_z/\partial y$, can be shown to be

$$P_{rad} = j(\delta/\omega\epsilon_0)|D|^2 K. \qquad (8.24)$$

Now since we are only interested in a perturbation solution, we can approximate δ by δ_0 and D by D_0 of the straight slab waveguide. The time-average power flow per unit length along the slab in this case is given by

$$P_T = \int_{-\infty}^{\infty} \tfrac{1}{2}\,\mathrm{Re}\,(\boldsymbol{E} \times \boldsymbol{H}^*)\cdot \boldsymbol{a}_s dy \qquad (8.25)$$

which again according to (8.11) and the expression $E_y = (k_{00}'/\omega\epsilon_0)(\epsilon_r^{-1}, 1)H_z$ for both inside and outside of the slab, can be shown to be

$$P_T = (k_{00}'/4\omega\epsilon_0)|D|^2\left[\frac{2}{\delta_0} + \frac{|B|^2}{\epsilon_r|D|^2}\left(b + \frac{\sin\gamma_0 b}{\gamma_0}\right)\right]$$

for a TM-mode symmetric in y. Thus with the use of the relationship $B = D\sec\psi_0$ where $\psi_0 = b\gamma_0/2$ and the secular equation $\tan\psi_0 = \epsilon_r\delta_0/\gamma_0$ in (8.18) we can further derive an explicit relation between $|D|$ and P_T as

$$|D| = \left[\frac{4\omega\epsilon_0 P_T \gamma_0^2 \epsilon_r}{k'_{00}(k'^2_{00} + \epsilon_r\delta_0^2)(\epsilon_r - 1)b_e}\right]^{1/2}$$

where

$$b_e = b + \frac{2}{\delta_0}\left(\frac{k^2}{k'^2_{00} + \epsilon_r\delta_0^2}\right) \tag{8.26}$$

The relative power loss per unit arclength, s, can then be obtained from (8.26) and (8.24). Now since this power loss is directly attributable to the radiation of the surface wave, we can therefore calculate the attenuation constant from the well-known relationship $\alpha = (P_{rad}/2P_T)$ as

$$\alpha = \frac{\epsilon_r}{\epsilon_r - 1}\left[\frac{j2\delta_0\gamma_0^2 K}{k'_{00}(k'^2_{00} + \epsilon_r\delta_0^2)b_e}\right] \tag{8.27}$$

Equation (8.27) can be shown to be identical to the one previously derived using the perturbation method. It is apparent that one merely needs to replace H_z by E_z and use the relation $\boldsymbol{H} = (j/\omega\mu_0)\nabla \times \boldsymbol{E}$ in order to extend the result to the TE-mode. Without much ado, we can write down the expression for α corresponding to an even TE-mode (symmetric in y) as

$$\alpha_{TE} = \left(\frac{\epsilon_r}{\epsilon_r - 1}\right)\frac{j2\delta_0\gamma_0^2 K}{k'_{00}k^2 b_e}; \qquad b_e = b + \frac{2}{\delta_0} \tag{8.28}$$

which is equivalent to replacing $(k'^2_{00} + \epsilon_r\delta_0^2)$ by k^2 in (8.27). It is noteworthy that the expression for the effective width b_e in (8.28) is the width of the slab plus twice the penetration depth of the surface wave field (defined as the distance over which the evanescent field outside the slab decays to e^{-1} of its value at the slab surface). The difference in b_e for a TM-mode and a TE-mode simply reflects the fact that different amounts of energy are stored in the vicinity of the slab for the two types of surface wave field, even though the decay rate is the same.

8.5 Asymmetric surface-wave modes

Up to now we have only discussed the radiation loss of a surface wave field which is symmetrical to the slab in the limit when the curvature vanishes. The first-order change in the propagation constant, i.e. B_1 as defined in (8.7), is exponentially small in that case and therefore can be ignored in the calculation of the attenuation constant which contains the factor B_1 implicitly in K. However, it remains to be seen whether this is still the case when the surface wave field is no longer symmetric. As we point out later (see section 9.6), a contribution of

$O(1)$ to the exponential in K exists if B_1 itself has a significant magnitude.

For convenience, we consider first the case of a TE-mode. Writing the solution in a form similar to (8.6), the transverse electric field component in this case is expressed by

$$E = E_0 + \chi E_1 + \ldots .$$
$$k' = k'_0(1 + \chi b B_1 + \ldots) \quad (8.29)$$

where $\chi = 1/R_0$ is the small expansion parameter. For the calculation of K, we are interested in that part of B_1 (if any) which is *not* exponentially small. This means we can replace k'_0 by k'_{00} and consider E_0 to be the surface wave field of a straight slab. Following the same procedure as given in sections 8.2 and 8.3, we can extract from the Helmholtz equation (8.9), the differential equations for the zero-th order and first-order solutions as follows

$$\partial^2 E_0/\partial y^2 + [k_0^2 \epsilon_r(y) - k_{00}'^2]\, E_0 = 0 \quad (8.30a)$$

$$\partial^2 E_1/\partial y^2 + [k_0^2 \epsilon_r(y) - k_{00}'^2]\, E_1 = -\partial E_0/\partial y + 2k_0'^2 E_0(2B_1 b - 2y) \quad (8.30b)$$

Here, we use the notation $\epsilon_r(y)$ to denote the relative permittivity in regions both inside and outside of the slab. (For a symmetrical slab, we therefore have $\epsilon_r(y) = 1$ for $|y| \geqslant b/2$.) Multiplying (8.30a) by E_1 and (8.30b) by E_0 and then integrating the difference of the equations from $-\infty$ to ∞, we find that

$$[E_0 \partial E_1/\partial y - E_1 \partial E_0/\partial y]_{y=-\infty}^{\infty} = -(E_0^2/2)|_{y=-\infty}^{\infty} + k_{00}'^2 \int_{-\infty}^{\infty} E_0^2(y)\,(bB_1 - 2y)dy$$

Thus, by requiring both E_0 and E_1 to vanish at $\pm\infty$, together with the boundary conditions at the slab surfaces, the following expression for B_1 is obtained

$$B_1 = \frac{\int_{-\infty}^{\infty} y E_0^2(y) dy}{b \int_{-\infty}^{\infty} E_0^2(y) dy} \quad (8.31)$$

Now since only the square of E_0 appears in both integrals, B_1 is identically zero for a symmetric as well as an antisymmetric mode of a symmetric slab waveguide. Consequently, the expression for K as given in (8.12) can be further reduced to $K = K_s$ where

$$K_s = -\tfrac{1}{2} j \exp\left[\delta_0 b - 2R_0\left(k' \log\frac{\delta_0 + k'_{00}}{k_0} - \delta_0\right)\right] \quad (8.32)$$

Equation (8.32) can then be used for the computation of the attenuation constant for both the symmetric and antisymmetric surface-wave fields. To extend the result to include the TM-mode, we apply the same procedure to the zero-th and first-order solutions of H, with the exception that we multiply (8.10) by $\epsilon_r^{-1}(y)H_1$, instead of H_1, and $\epsilon_r^{-1}(y)H_0$ to (8.24). After a subtraction of the two equations and integration between $\pm\infty$, the result is given as follows

$$B_1 = \frac{\int_{-\infty}^{\infty}\epsilon_r^{-1}yH_0^2(y)dy + (\epsilon_r^{-1} - 1)[H_0^2(b/2) - H_0^2(-b/2)]/k_{00}'^2}{b\int_{-\infty}^{\infty}\epsilon_r^{-1}H_0^2(y)dy} \tag{8.33}$$

Thus, B_1 again vanishes for both the symmetric and antisymmetric modes of a symmetric slab waveguide. Since the exponentially small quantity has been consistently omitted, (8.33) therefore agrees with the more detailed derivation of B_1 given earlier in (8.21).

Obviously (8.31) and (8.33) can also be applied to the case where the surface wave field is asymmetrical due to different media appearing on the two outer sides of the dielectric slab. Denoting the permittivity of the medium interior to the slab as ϵ_{r2} so that $k_2 = k_0\,\epsilon_{r2}^{1/2}$ and $\delta_2 = (k_{00}'^2 - k_2^2)^{1/2}$, the TE-solution for such a waveguide can be shown to be

$$E_0 = \begin{cases} Ae^{\delta_2(y+b/2)}, & y<-b/2 \\ B\cos\gamma_0 y + C\sin\gamma_0 y, & -b/2<y<b/2 \\ De^{-\delta_0(y-b/2)}, & y>b/2 \end{cases} \tag{8.34}$$

with the secular equation given as

$$(\Delta - \tan\psi_0)(1 + \Delta_2\tan\psi_0) + (\Delta_2 - \tan\psi_0)(1 + \Delta\tan\psi_0) = 0 \tag{8.35}$$

where $\Delta = \delta_0/\gamma_0$, $\Delta_2 = \delta_2/\gamma_0$ and $\psi_0 = \gamma_0 b/2$. In addition, by applying the boundary conditions at each slab surface and by the use of (8.35), we can also prove that

$$\begin{aligned} B^2 + C^2 &= (1+\Delta^2)D^2 = (1+\Delta_2^2)A^2 \\ (2\sin 2\psi_0)BC &= (1-\Delta^2)D^2 = (1-\Delta_2^2)A^2 \\ \cos\psi_0 &= \pm(\gamma_0^2 - \delta_2\delta_0)/\Omega \\ \sin\psi_0 &= \pm\gamma_0(\delta_0 + \delta_2)/\Omega \\ \Omega &= [(\gamma_0^2 + \delta_0^2)(\gamma_0^2 + \delta_2^2)]^{1/2} \end{aligned}$$

The sign of square root must be chosen the same for both $\sin\psi_0$ and $\cos\psi_0$ in order to represent a given mode of the slab guide. After considerable algebra, we then arrive at a surprisingly simple expression for the two integrals in (8.31) as[3,4]

$$\begin{aligned} \int_{-\infty}^{\infty} y E_0^2(y) dy &= \frac{D^2 (k^2 - k_0^2) \hat{b}_e}{4\gamma_0^2} \left(\frac{1}{\delta_0} - \frac{1}{\delta_2} \right) \\ \int_{-\infty}^{\infty} E_0^2(y) dy &= \frac{D^2 \hat{b}_e (k^2 - k_0^2)}{2\gamma_0^2} \end{aligned} \tag{8.36}$$

where the effective width $\hat{b}_e$ for the asymmetrical slab is defined as

$$\hat{b}_e = b + \frac{1}{\delta_0} + \frac{1}{\delta_2}$$

Substitution of these results into (8.26) yields a finite value for B_1 as

$$B_1 = \frac{1}{2b} \left(\frac{1}{\delta_0} - \frac{1}{\delta_2} \right)$$

Clearly B_1 can be significant, if the penetration depth δ_0^{-1} of the surface wave field into the exterior region ($y > b/2$) differs substantially from the penetration depth δ_2^{-1} into the interior region ($y < -b/2$). As a result, the expression for K for an asymmetric slab is now given as

$$K = K_s \exp \left\{ - \left(\frac{1}{\delta_0} - \frac{1}{\delta_2} \right) k'_{00} \log \frac{\delta_0 + k'_{00}}{k_0} \right\} \tag{8.37}$$

where K_s as given in (8.27) is the corresponding value of K in the symmetric slab.

As mentioned in Chapter 7, a case of interest in fiber optical waveguide applications is when $(k - k_0)$ and $(k_2 - k_0)$ are much smaller than k_0. Thus, $k'_{00} \approx k_0$ and $\delta_0 \ll k_0$. Expansion of the logarithmic function further reduces (8.37) to

$$K \simeq K_s \exp [-(1 - \delta_0/\delta_2)]$$

Aside from the difference in K, the power carried by the surface wave field of an asymmetric slab is also different. More specifically, with the use of the Poynting vector, we have according to (8.36)

$$\begin{aligned} P_{TE} &= \left(\frac{k'_{00}}{2\omega\mu_0} \right) \int_{-\infty}^{\infty} E_0^2(y) dy \\ &= \frac{D^2 \hat{b}_e (k^2 - k_0^2) k'_{00}}{4\omega\mu_0 \gamma_0^2} \end{aligned}$$

Thus by finding the power radiated P_{rad} for the TE mode, we can obtain the attenuation constant as

$$\alpha_{TE} = \left(\frac{\epsilon_r}{\epsilon_r - 1}\right) \frac{j2\delta_0\gamma_0^2 K_s}{k'_{00}k^2\hat{b}_e} \exp\left(\frac{\delta_0}{\delta_2} - 1\right) \tag{8.38}$$

which has the identical form as in (8.28) for a symmetrical slab, except for an additional factor, $\exp(\delta_0/\delta_2 - 1)$. Depending on the ratio of penetration depths, the radiation loss on a curved slab therefore can be either suppressed or enhanced.

8.6 Relation to stored energy

Returning to the formal solution in (8.31) for a TE-mode, it is of interest to note that the first-order change in the propagation constant of a curved slab can be also interpreted from the viewpoint of change in stored energy from a straight to a curved slab. First of all, the integral in the numerator of (8.31) can be rearranged using (8.30a) and integration-by-parts as follows

$$2k_{00}'^2 \int_{-\infty}^{\infty} yE_0^2(y)dy = \int_{-\infty}^{\infty} y\,\{E_0^2(y)[k_{00}'^2 + k_0^2\epsilon_r(y)] - [\partial E_0(y)/\partial y]^2\}dy$$

To obtain the above expression use has also been made of the exponential decay of E_0 at $y = \pm\infty$. From the relationship that $H_{y0} = (k'_{00}/\omega\mu_0)E_0$ and $H_{s0} = (j\omega\mu_0)^{-1}\partial E_0/\partial y$, we then arrive at the following identity

$$\int_{-\infty}^{\infty} yE_0^2(y)dy = \left(\frac{\omega\mu_0}{k_{00}'^2}\right) \int_{-\infty}^{\infty} yW_e(y)dy \tag{8.39}$$

where

$$W_e(y) = \omega\epsilon \boldsymbol{E}_0 \cdot \boldsymbol{E}_0/2 + \omega\mu_0 \boldsymbol{H}_0 \cdot \boldsymbol{H}_0/2 \tag{8.40}$$

Now since $\boldsymbol{E}_0$ and $\boldsymbol{H}_0$ are the electric and magnetic field vectors of a straight slab, the expression $W_e(y)$ defined in (8.40) corresponds exactly to the stored energy density at a given point in space. Following a similar derivation, the integral appearing in the denominator of (8.31) can also be expressed in terms of W_e as

$$\int_{-\infty}^{\infty} E_0^2(y)dy = \left(\frac{\omega\mu_0}{k_{00}'^2}\right) \int_{-\infty}^{\infty} W_e(y)dy \tag{8.41}$$

Consequently, the first-order change in the propagation constant from a curved to a straight slab is now explicitly related to the stored energy density according to (8.7) and (8.31) as

$$\frac{k' - k'_{00}}{k'_0} = \frac{\dfrac{1}{R_0}\displaystyle\int_{-\infty}^{\infty} \rho\, W_e(y)dy}{\int_{-\infty}^{\infty} W_e(y)dy} - 1 \tag{8.42}$$

where $\rho = y + R_0$ is the radial distance of the global co-ordinate system describing the curved slab. If now it is assumed that the surface wave field of a curved slab is the same as the corresponding straight slab in a cross-section given approximately by several effective widths, we can replace the integral in the numerator by

$$V_e = \int_0^\infty W_e(\rho)\rho d\rho = \frac{1}{\Delta\phi}\int_S W_e dS \tag{8.43}$$

which then represents the energy per unit length of the surface-wave field stored in a thin sectoral volume S defined by $0 < \rho < \infty$ and $\phi_0 < \phi < \phi_0 + \Delta\phi$ as $\Delta\phi \to 0$. (Fig. 8.2*b*). On the other hand the integral in the denominator is related to the energy per unit length stored in a thin slice of volume S_0 defined by $-\infty < y < \infty$ and $s_0 < s < s_0 + \Delta s$ where $\Delta s = R_0 \Delta\phi$ (Fig. 8.2*a*). In the corresponding case of a straight slab

$$V_{e0} = R_0 \int_{-\infty}^{\infty} W_e(y) dy = \frac{1}{\Delta\phi}\int_{S_0} W_e dS \tag{8.44}$$

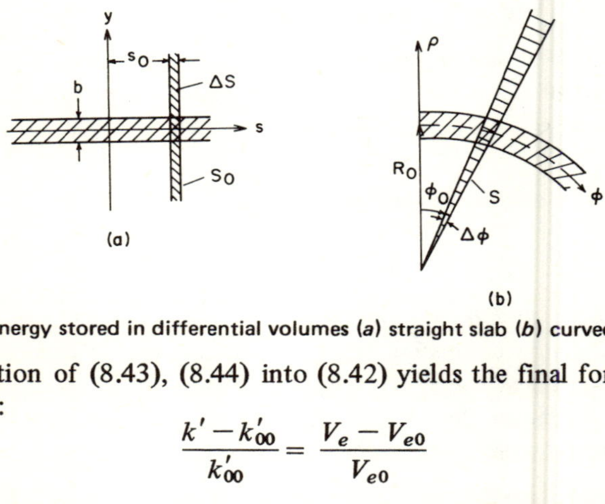

Fig. 8.2 Energy stored in differential volumes (*a*) straight slab (*b*) curved slab

Substitution of (8.43), (8.44) into (8.42) yields the final form of the theorem:

$$\frac{k' - k'_{00}}{k'_{00}} = \frac{V_e - V_{e0}}{V_{e0}} \tag{8.45}$$

Accordingly, the percentage change in the propagation constant from a curved to a straight slab is directly proportional to the percentage change in the stored energy contained in the two corresponding volumes.

8.7 Radiation from discontinuities in curvature

The imaginary part of k', calculated from (8.23), gives the attenuation due to radiation caused by the bend. The relevant contribution is due to the factor K which contains an exponentially decreasing term. At a discontinuity in curvature, radiation may be expected having an inverse radius of curvature dependence; and depending on the numerical factors involved the one source of radiation may be dominant over the other. Clearly for small enough curvatures it is the discontinuity effect that should dominate.

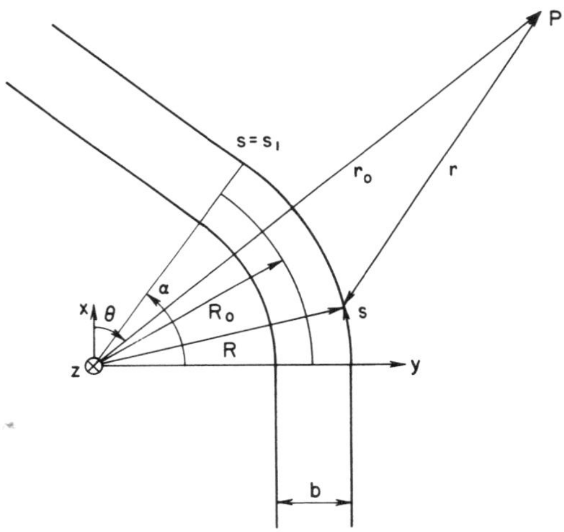

Fig. 8.3 Curved dielectric slab between straight sections

The junction of a straight to curved section of a dielectric slab involves a discontinuity of curvature which of itself is a potential cause of radiation. In addition the discontinuity will disturb the field matching at the junction and will be responsible for both a reflected wave and the generation of higher-order modes. These will cause additional radiation effects, all proportional to the curvature. The calculation of this latter radiation is extremely difficult, and is considered later in Chapter 11. Here we shall be concerned with the first effect only, and will follow in general terms the derivation of Maley[5] *et al.* who considered a TE mode in a bend of angle α set between two infinite straight sections.

Fig. 8.3 shows the geometry of the arrangement. Only the radiation from the outer surface is considered, and an incident wave in the slab from $x = -\infty$ encounters the bend at $x = 0$. It is convenient to consider a single variable s *on the outer surface* so s and x coincide from $-\infty$ to 0. At the second junction s takes the value $s_1 = R\alpha$ where R

is the radius of curvature to the outside surface. If R_0 is the radius of curvature of the slab centerline then

$$R = R_0 + \tfrac{1}{2}b \tag{8.46}$$

On the slab the wave propagates as $\exp(-jk_1 s)$ where $k_1 = k'_{00}$ on the straight sections.* Relative to the centerline, (8.23) gives the value of k_1 in the curved section. Since we are here concerned with the discontinuity effect the extra (exponentially small) difference between k' and k'_{00} will be ignored. However since k' refers to the centerline and k_1 refers to the outside surface we must have $k_1 R = k'R_0$ on the curved section, or

$$k_1 = k'R_0/(R_0 + \tfrac{1}{2}b) \tag{8.47}$$

At $s_1 = \alpha R$ the propagation constant reverts from $k'R_0/(R_0 + \frac{1}{2}b)$ to k'_{00}. Hence an extra phase of amount $\alpha k'b$ has to be added for $s > s_1$ to ensure phase continuity at $s = s_1$.

It is convenient to define

$$\begin{aligned} k' \approx k'_{00} &= pk_0 \\ k'R_0/(R_0 + \tfrac{1}{2}b) &= p_1 k_0 \end{aligned} \tag{8.48}$$

so that $p > 1, p_1 > 1$ and

$$p_1 = p/(1 + \tfrac{1}{2}b/R_0) \approx p(1 - \tfrac{1}{2}b/R_0) \quad \text{for} \quad R_0 \gg b \tag{8.49}$$

So far as orders of magnitude are concerned the inequality $R_0 \gg b$ stipulates a large radius of curvature relative to the slab thickness and also, by implication, relative to the free-space wavelength. However, the far-field pattern at distance r is taken at still greater distances with $r \gg R_0$. And finally the integration limits on the straight sections are taken to infinity so that the limits on s are $\gg r$.

The field at a point P can be formulated through Schelkunoff's induction theorem[6] in terms of fictitious electric and magnetic current sheets $\boldsymbol{J}$ and $\boldsymbol{M}$ on the outer surface of the slab. If $\boldsymbol{H}_t$ and $\boldsymbol{E}_t$ are respectively the tangential components of the magnetic and electric fields at the slab surface and $\boldsymbol{a}_n$ the unit outward normal, then

$$\boldsymbol{J} = \boldsymbol{a}_n \times \boldsymbol{H}_t, \quad \boldsymbol{M} = -\boldsymbol{a}_n \times \boldsymbol{E}_t \tag{8.50}$$

For the slab mode $\boldsymbol{H}_t$ is directed along the s-co-ordinate, so $\boldsymbol{a}_n \times \boldsymbol{H}_t = -\boldsymbol{a}_z H$. Hence the magnetic vector potential $\boldsymbol{A}$ is given by

* The value of k'_{00} for the TE mode differs from that given by (8.15) and is found later in this section

$$A = \frac{-a_z H}{4\pi} \int_{-\infty}^{\infty} \int_{-\infty}^{\infty} \frac{e^{-jk_1 s}\, e^{-jk_0 \rho}}{\rho}\, dz' ds \tag{8.51}$$

where

$$\rho^2 = (z - z')^2 + r^2$$

and r is the normal distance from a point on the slab surface to P, as indicated in Fig. 8.3.

In (8.51) the z' integration can be carried out to give

$$A = (j/4) a_z H \int_{-\infty}^{\infty} e^{-jk_1 s} H_0^{(2)}(k_0 r) ds \tag{8.52}$$

For all values of s, $k_0 r \gg 1$ so that the asymptotic form of the Hankel function can be used, giving

$$A \sim -(a_z/4) H e^{-j\pi/4} (2/\pi k_0)^{1/2} \int_{-\infty}^{\infty} e^{-j(k_1 s + k_0 r)} r^{-1/2} ds \tag{8.53}$$

By introducing a factor $\partial(k_1 s + k_0 r)\partial s$ in the numerator and denominator of the integrand, (8.53) can be integrated by parts, and the process repeated as often as necessary. Provided that the derivatives of $k_1 s + k_0 r$ stay finite and the first derivative does not go to zero anywhere, all such terms integrate to zero between the infinite limits. Since $\partial r/\partial s$ is limited to ± 1 at end-fire positions, and $k_1 > k_0$ it is clear that the first derivative is never zero. *Hence it is only the discontinuities in k_1 that give rise to terms leading to radiation.* Breaking the range into $(-\infty, 0)$, $(0, s_1)$ and (s_1, ∞) and integrating by parts (8.53) gives

$$A = -(a_z/4k_0) H e^{j\pi/4} (2/\pi k_0)^{1/2} \left\{ \left. \frac{e^{-jk_0(ps + r - p\alpha b)}}{(p + \partial r/\partial s) r^{1/2}} \right|_{s_1}^{\infty} + \left. \frac{e^{-jk_0(p_1 s + r)}}{(p_1 + \partial r/\partial s) r^{1/2}} \right|_0^{s_1} + \left. \frac{e^{-jk_0(ps + r)}}{(p + \partial r/\partial s) r^{1/2}} \right|_{-\infty}^{0} \right\} \tag{8.54}$$

+ higher order terms for large r.

It is clear that but for the change from p to p_1 in the curved section, there would be no radiation to this order. If, as in Fig. 8.3, θ is the angle from the x – axis to the field point, then $\partial r/\partial s = -\cos\theta$ at $s = 0$ and $-\cos(\theta + \alpha)$ at $s = s_1$, for $r \gg R_0$. This gives the form of A for $r \gg R$.

$$A \sim -\frac{a_z H\, e^{j\pi/4} e^{-jk_0 r_0}}{4k_0 (\pi k_0 r_0/2)^{1/2}} \left[e^{jk_0 R \sin\theta} f(\theta) - e^{jk_0 [R \sin(\theta + \alpha) - p_1 R\alpha]} f(\theta + \alpha) \right] \tag{8.55}$$

where

$$f(\theta) = \frac{1}{p - \cos\theta} - \frac{1}{p_1 - \cos\theta} \tag{8.56}$$

and r_0 is the distance from P to the center of curvature of the bend. The magnetic current sheet gives rise to an electric vector potential $\boldsymbol{F}$. Since $\boldsymbol{E}_t$ is directed along the z-axis, $-\boldsymbol{a}_n \times \boldsymbol{E}_t = -\boldsymbol{a}_s E$. Hence

$$\boldsymbol{F} = -\frac{E}{4\pi}\int_{-\infty}^{\infty}\int_{-\infty}^{\infty} \boldsymbol{a}_s \frac{e^{-jk_1 s}e^{-jk_0\rho}}{\rho}\, dz'ds \tag{8.57}$$

Using almost identical procedures as used for A the integrations for $\boldsymbol{F}$ may be performed, and for large r give

$$\boldsymbol{F} \sim -\frac{Ee^{j\pi/4}e^{-jk_0r_0}}{4k_0(\pi k_0 r_0/2)^{1/2}}\{\boldsymbol{a}_y \sin\alpha\, e^{jk_0[R\sin(\theta+\alpha)-p_1R\alpha]}f(\theta+\alpha)$$

$$-\boldsymbol{a}_x\,[\,e^{jk_0R\sin\theta}f(\theta) - \cos\alpha e^{jk_0[R\sin(\theta+\alpha)-p_1R\alpha]}\,f(\theta+\alpha)]\} \tag{8.58}$$

From (8.55) and (8.58) the radiated electric far-field can be obtained from

$$\boldsymbol{E} = -j\omega\mu_0 A - \text{curl}\,\boldsymbol{F} \tag{8.59}$$

which gives

$$\boldsymbol{E} = \frac{j\boldsymbol{a}_z e^{j\pi/4}e^{-jk_0r_0}E}{4(\pi k_0 r_0/2)^{1/2}}[(Y-\sin\theta)e^{jk_0R\sin\theta}f(\theta)$$

$$-[Y-\sin(\theta+\alpha)]e^{jk_0[R\sin(\theta+\alpha)-p_1R\alpha]}f(\theta+\alpha)] \tag{8.60}$$

Here, $Y = (\mu_0/\epsilon_0)^{1/2}\,H/E$ is the normalized admittance of the mode at the slab surface. It is obtainable from an analysis similar to that of section 8.2. For $K = 0$ we take

$$E_z = e^{-jk's}E \tag{8.61}$$

where

$$E = E_0 + \chi E_1 + \ldots \tag{8.62}$$

$$k' = k_0'(1 + \chi b B_1 + \ldots) \tag{8.63}$$

and define quantities γ, δ and ψ as in (8.8).

Then E_0 will satisfy the same equation as H_0, namely (8.10), and gives, for a symmetrical mode,

$$E_0 = \begin{cases} A\cos\gamma y & 0<y<b/2 \\ A\cos\psi e^{-\delta(y-b/2)}, & y>b/2 \end{cases} \tag{8.64}$$

From $\text{curl}\,E = -j\omega\mu_0 H$ we get $-j\omega\mu_0 H_x = \partial E_z/\partial y$ or $H_x = e^{-jk's}H$ where

$$H = \begin{cases} -(j/\omega\mu_0)A\gamma\sin\gamma y, & 0<y<b/2 \\ -(j/\omega\mu_0)A\delta\cos\psi e^{-\delta(y-b/2)}, & y>b/2 \end{cases} \tag{8.65}$$

Matching H_x at $y = b/2$ gives the secular equation

$$\tan \psi = \delta/\gamma \tag{8.66}$$

corresponding to (8.18) in the TM case.

From (8.64) and (8.65) we get the ratio $Y = (\mu_0/\epsilon_0)^{1/2} H/E$ at the slab surface

$$Y = -j\delta/k_0 \tag{8.67}$$

Since $k_0' > k_0$, δ is real and Y is imaginary (for a lossless dielectric slab).

8.8 Radiation pattern and radiated power

From (8.60) it is seen that the radiated field is in the form of an interfererence pattern from two sources located at the two discontinuities. It consists of a number of lobes of slowly varying widths and amplitudes, as sketched in Fig. 8.4. The widest lobe is deflected from the vertical in the same direction as the bend, but through an angle only half as great, i.e. $\alpha/2$.

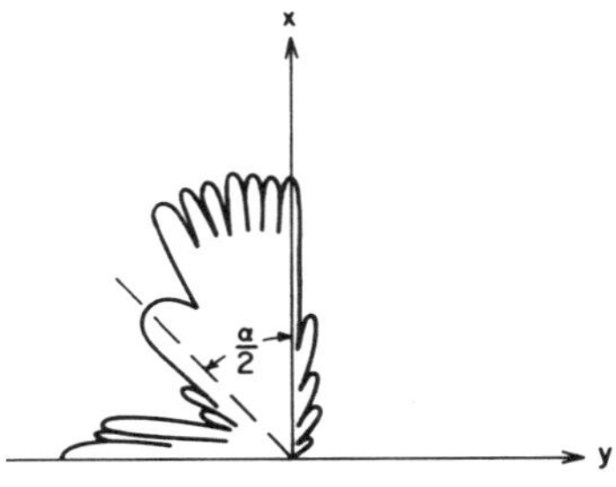

Fig. 8.4 Field strength radiation pattern for 90° bend

Inserting in (8.60) the value $\cos\psi$ for E at the slab surface, using (8.67) for Y, and making the approximation, for p_1 nearly equal to p, of

$$f(\theta) \approx -(\tfrac{1}{2} bp/R)(p - \cos\theta)^{-2} \tag{8.68}$$

gives the following relation for the power density pattern from a single discontinuity

$$|E_z|^2 = \frac{A^2 \cos^2 \psi}{8\pi k_0 r_0} (\sin^2\theta + \delta^2/k_0^2) f^2(\theta) \tag{8.69}$$

The justification for taking the power density from only a single discontinuity lies in the nature of the mutual interference phenomena: when two sources are well separated in wavelengths so as to give a many-lobed pattern the power radiated is very nearly the double of that from a single source. Obviously (8.69) would not be usable if the discontinuities were close.

An integration and doubling of (8.69) gives the power per unit width radiated by the discontinuities. Maley[7] gives the rather lengthy exact expression when the interference is not neglected. Working with (8.69) gives

$$P_r = \frac{(k'b\cos\psi/4\pi k_0 R)^2 A^2}{120k_0}\int_{-\alpha}^{\pi}\frac{\sin^2\theta + \delta^2/k_0^2}{(p-\cos\theta)^4}\,d\theta \tag{8.70}$$

Since $p^2 - 1 = \delta^2/k_0^2$ the integral can be expressed in the form

$$\int_{-\alpha}^{\pi}\frac{p^2-\cos^2\theta}{(p-\cos\theta)^4}\,d\theta = \frac{d}{dp}\cdot p\frac{d}{dp}\int_{-\alpha}^{\pi}\frac{d\theta}{p-\cos\theta} = \frac{d}{dp}\cdot p\frac{d}{dp}\left\{\frac{\pi + \tan^{-1}\dfrac{(p^2-1)^{1/2}\sin\alpha}{p\cos\alpha - 1}}{(p^2-1)^{1/2}}\right\}$$

On carrying out the differentiation, replacing p by k'/k_0 and insertion into (8.70) we get

$$P_r = \left(\frac{k'b\cos\psi}{4\pi k_0 R}\right)^2\frac{A^2 k_0}{120\delta^5}\left\{\frac{\delta k_0(2k'^2 + k_0^2)\sin\alpha}{k' - k_0\cos\alpha} + \frac{\delta^3 k' k_0\sin\alpha}{(k' - k_0\cos\alpha)^2} + k'(k'^2 + 2k_0^2).\left[\pi + \tan^{-1}\frac{\delta\sin\alpha}{k'\cos\alpha - k_0}\right]\right\} \tag{8.71}$$

This expression needs to be related to the relative power lost at the two discontinuities. It is therefore necessary to refer the amplitude A to the power flowing per unit width in the slab. This is given by integrating the square of the quantities in (8.64), with an additional factor k'/k_0 in the dielectric,

$$\begin{aligned} P_0 &= \frac{2A^2}{240\pi}\left[(k'/k_0)\int_0^{b/2}\cos^2\gamma y\,dy + \int_{b/2}^{\infty}\cos^2\psi\, e^{-2\delta(y-b/2)}dy\right] \\ &= \frac{A^2}{240\pi}\left[\frac{k'}{k_0}\left(\frac{b}{2} + \frac{\sin^2\psi}{\delta}\right) + \frac{\cos^2\psi}{\delta}\right] \end{aligned} \tag{8.72}$$

Hence, on eliminating A between (8.71) and (8.72),

$$\frac{P_r}{P_0} = \frac{(k'\gamma b/R\delta^2)^2/8\pi}{k_0\gamma^2 + k'\delta[\delta + \frac{1}{2}b(\gamma^2+\delta^2)]}$$
$$\left\{\frac{[3k'^3 - k_0(k_0^2 + 2k'^2)\cos\alpha]\delta k_0 \sin\alpha}{(k' - k_0\cos\alpha)^2} + k'(k'^2 + 2k_0^2)\cdot\left[\pi + \tan^{-1}\frac{\delta\sin\alpha}{k'\cos\alpha - k_0}\right]\right\} \tag{8.73}$$

In this expression k', γ and δ are defined in (8.8), and γ satisfies the secular equation (8.66) with $\psi = b\gamma/2$. No distinction has been made here between k' and k_0', since the attenuation round the bend, which depends exponentially on R, has been assumed negligible in calculating the discontinuity effects. As given by (8.73), these vary as R^{-2}.

The above calculations give the radiation on the outside of the bend. On the inside a comparable calculation can be made, and Maley (private communication) has indicated that the radiation is similar though somewhat smaller. However, on the inside there are further complications. The concave surface will support whispering-gallery modes, whilst incident radiation from one straight section can be reflected from the other. An attempt to evaluate these inter-related aspects is covered in Chapter 11.

References

1 Marcatili, E.A.J.: 'Bends in optical dielectric guides', *Bell System Technical Journal,* Sept. 1969, pp. 2103–2132

2 Dang, N.D.: Dept. Electr. Engr., Queen Mary College, London

3 Marcuse, D.: 'Bending losses of the asymmetric slab waveguide,' *Bell System Technical Journal*, Oct. 1971, pp. 2551–2563

4 Kuester, E.F., and Chang, D.C.: 'Surface-wave radiation loss from curved dielectric slabs and fibers,' *IEEE J. Quantum Electron.*, Nov. 1975, pp. 903–907

5 Maley, S.W., Chang, D.C., and Haddad, H.: 'Radiation from a bend in a dielectric slab waveguide'. USNC-URSI Conference, Program and Abstracts, Boulder, 1974, paper VI 10–7, p. 178

6 Schelkunoff, S.A.: 'Electromagnetic Waves' (Van Nostrand, New York, 1943)

7 Maley, S.W.: 'Radiation from a circular bend between two discontinuities in a dielectric slab waveguide', Scientific Report No. 10, Electromagnetics Laboratory, University of Colorado, Boulder, Colorado, 1974

Chapter 9

The curved dielectric fiber

9.1 Application of the radiation condition

The analysis for a straight fiber has been given by, among others, Kao[1] and Clarricoats.[2] A cental core, of size of the order of a few wavelengths, is surrounded by a cylindrical cladding of much greater diameter. The difference in their dielectric constants is small, typically of the order of 1% or so. The field in the cladding is attenuated approximately exponentially away from the core and is negligibly small at the outer surface. Although a refined calculation of the radiation due to bending would take into account both the core-cladding and the cladding-air interfaces, it seems probable that the major effect would occur at the former because of the large attenuation of the field at the surface of the cladding. We therefore consider the cladding diameter infinite, and seek to apply the radiation condition (7.32) directly at the surface of the core. It is apparent that there are two complications to this process; (i) the radiation condition is polarization sensitive, and (ii) there is a train of angular components generated along with the incoming form of the exciting component. All these radiation terms are of order σ, where σ is small and is given by (7.31). Corresponding to a typical term $I_n \cos n\psi$ outside the core, there will be a like term inside the core, to ensure continuity of tangential electric fields. But because of the difference in the core and cladding dielectric constants the tangential magnetic fields will not be quite matched, and an additional term in $K_n \cos n\psi$ will hence be required in the cladding. Because of the radiation condition to be applied to *this* term a new train of angular components, of order σ^2 will now be generated; and so on. With σ small enough these higher order terms can be neglected. The total effect on the exciting mode can therefore be ignored, and only that component with the same angular variation as the exciting mode need be retained for consideration.

Because of the form taken by (7.32) it is necessary to attend to the angular variation and the resulting plane of polarization. Remembering that E_ϕ in Fig. 7.2 corresponds to the axial direction s, we consider a field E_s proportional to $\cos n(\psi + \alpha)$, where α is an arbitrary angle determining the plane of polarization. It is convenient here to work with complex forms, and since j is already used for the time variable we go to bi-complex forms with i, as in Chapter 1, a second complex variable: $i^2 = -1$, $j^2 = -1$, but $ij \neq -1$. All operations in this section of taking the real or imaginary parts refer to the field of i. Hence we use

$$\cos m(\psi + \alpha) = \operatorname{Re} e^{-im(\psi + \alpha)} \tag{9.1}$$

The exciting mode form in the cladding of a straight fiber can be taken proportional to $e^{-j\nu\phi} K_m [k\rho(\gamma^2 - 1)^{1/2}] \cos m(\psi + \alpha)$ and will therefore be replaced by $e^{-im\alpha} K_m$, with an implied factor $e^{-im\psi - j\nu\phi}$. However, because of the form of (7.32) we need first to write $\cos m(\psi + \alpha)$ as $\cos m\psi \cos m\alpha - \sin m\psi \sin m\alpha$ and apply the radiation contition to each term separately. To first order there is coupling only to the cosine term, and apart from the neglected entrained modes this will introduce a radiation term $I_m \cos m\psi \cos m\alpha$. More explicitly, we can put the relevant field terms in the form

$$\begin{aligned} E_s = {} & e^{-j\nu\phi} \{K_m (\cos m\psi \cos m\alpha - \sin m\psi \sin m\alpha) \\ & + \sigma\epsilon_m I_m \cos m\psi \cos m\alpha\} \end{aligned} \tag{9.2}$$

(where the argument of the Bessel functions is $k\rho(\gamma^2 - 1)^{1/2}$). In complex form the right-hand side of (9.2) would be written as $(K_m + \sigma\epsilon_m e^{im\alpha} \cos m\alpha I_m) e^{-im\alpha - j\nu\phi}$. Thus if we define $\delta_{m\alpha}$ by

$$\delta_{m\alpha} = \sigma\epsilon_m e^{im\alpha} \cos m\alpha \tag{9.3}$$

we can treat $\delta_{m\alpha}$ as the mode self-coupling coefficient for radiation. It is implied, in this way of treating the problem, that the as yet unspecified amplitude of K_m apart from the above factor $e^{-im\alpha}$, is real in i. Thus (9.2), via α, determines the orientation of the field relative to the plane of bending. The same interpretation applies, *mutatis mutandis*, inside the fiber core.

9.2 HE mode in curved fiber

Following Clarricoats[2] for the field in the straight fiber, we can take, for the fields inside the core,

$$\begin{aligned} E_s &= a_m e^{-im\alpha} J_m \\ H_s &= b_m e^{-im\beta} J_m \end{aligned} \tag{9.4}$$

with an implied factor $e^{-im\psi - jk's}$. Both the as yet undetermined amplitudes a_m and b_m must be real (in i). The polarization angle β of the H_s component has to be determined in relation to α; it is not arbitrary. The argument of the Bessel functions is $K\rho$ where

$$\begin{aligned} K &= (\omega^2 \epsilon\mu_0 - k'^2)^{1/2} \\ \epsilon &= \text{core permittivity} \\ k' &= \text{axial propagation constant} \end{aligned} \tag{9.5}$$

Two independent transverse components are needed for field matching, and these can be conveniently taken as H_ψ and H_ρ:

$$H_\psi = \frac{ijk'm}{K^2\rho} b_m J_m e^{-im\beta} - \frac{j\omega\epsilon}{K} a_m J'_m e^{-im\alpha} \tag{9.6}$$

$$H_\rho = \frac{-jk'}{K} b_m J'_m e^{-im\beta} - \frac{ijm\omega\epsilon}{K^2\rho} a_m J_m e^{-im\alpha} \tag{9.7}$$

Outside the core, analagously to (9.4), (9.6) and (9.7) we take

$$\begin{aligned} E_s &= A_m (K_m + \delta_{m\alpha} I_m) e^{-im\alpha} \\ H_s &= B_m (K_m + \delta_{m\beta} I_m) e^{-im\beta} \end{aligned} \tag{9.8}$$

$$\begin{aligned} H_\psi = &- \frac{ijk'm}{K_c^2\rho} B_m (K_m + \delta_{m\beta} I_m) e^{-im\beta} \\ &+ \frac{j\omega\epsilon_c}{K_c^2} A_m (K'_m + \delta_{m\alpha} I'_m) e^{-im\alpha} \end{aligned} \tag{9.9}$$

$$\begin{aligned} H_\rho = &\frac{jk'}{K_c} B_m (K'_m + \delta_{m\beta} I'_m) e^{-im\beta} \\ &+ \frac{ij\omega\epsilon_c}{K_c^2\rho} A_m (K_m + \delta_{m\alpha} I_m) e^{-im\alpha} \end{aligned} \tag{9.10}$$

A_m, B_m are real (in i), the argument of the Bessel functions is $K_c\rho$ where the subscript c refers to the cladding, and

$$
\begin{aligned}
K_c &= (k'^2 - k_c^2)^{1/2} \\
k_c^2 &= \omega^2 \epsilon_c \mu_0 \qquad (9.11) \\
\epsilon_c &= \text{cladding permittivity}
\end{aligned}
$$

In (9.8) the term in $\delta_{m\alpha}$ and $\delta_{m\beta}$ account for the radiation effects due to the curvature.

We define the following quantities

$$
\begin{aligned}
x &= Ka, \quad a = \text{core radius} \\
y &= K_c a \qquad (9.12) \\
F_m(x) &= xJ'_m(x)/J_m(x)
\end{aligned}
$$

In addition we need the functions $M_m(y)$ and $\bar{M}_m(y)$ where

$$
\begin{aligned}
M_m(y) &= y\,\frac{K'_m(y) + \delta_{m\alpha} I'_m(y)}{K_m(y) + \delta_{m\alpha} I_m(y)} \qquad (9.13) \\
&\approx \frac{yK'_m(y)}{K_m(y)}\left[1 + \frac{\delta_{m\alpha}}{yK_m(y)K'_m(y)}\right] \quad \text{for small} \quad \delta_{m\alpha}
\end{aligned}
$$

and $\bar{M}$ is M with $\delta_{m\alpha}$ replaced by $\delta_{m\beta}$.

9.3 Secular equation

At $\rho = a$ we require continuity of E_s, H_s, H_ρ and H_ψ. On eliminating the amplitudes A_m and B_m this gives the pair of equations

$$k'm\Lambda(x^2 + y^2) = \omega(\epsilon y^2 F_m + \epsilon_c x^2 M_m) \qquad (9.14)$$

$$k'\Lambda(y^2 F_m + x^2 \bar{M}_m) = (\epsilon y^2 + \epsilon_c x^2)m\omega \qquad (9.15)$$

with

$$\Lambda = (b_m/a_m)e^{i(m\alpha - m\beta + \pi/2)} \qquad (9.16)$$

Before eliminating Λ between (9.14) and (9.15) we note that, since a_m and b_m are real (in i) and the equations require Λ also to be real, it follows that β is related to α by

$$m\beta = m\alpha + (n + \tfrac{1}{2})\pi, \; n \text{ any integer.} \qquad (9.17)$$

From this we can readily verify the important equation

$$\delta_{m\alpha} + \delta_{m\beta} = \epsilon_m \sigma \qquad (9.18)$$

a result that is independent of the field orientation angle α. This deduction would be strictly true if M_m and $\bar{M}_m$ were exactly real, and thus would be so if M_m did not contain terms in $\delta_{m\alpha}$, a small quantity of the order of σ. Hence we would contemplate β differing from its value given by (9.17) by a term of order σ, and (9.18) would then require an additional term of order σ^2. Hence (9.18) is correct to $0(\sigma)$, whilst $0(\sigma^2)$ terms have, in any case, been ignored in this analysis.

With β determined we can now proceed to eliminate Λ from (9.14) and (9.15). The resulting equation is rather complicated, but a great simplification is possible when ϵ is close to ϵ_c. (see Appendix F.) In this case the equation can be approximated by

$$[F_m(x) + x^2 N_m(y)]^2 = m^2(1 + x^2/y^2)^2 \tag{9.19}$$

where

$$N_m(y) = \frac{K'_m(y)}{yK_m(y)}\left[1 + \frac{\epsilon_m \sigma}{2yK_m(y)K'_m(y)}\right] \tag{9.20}$$

On taking the square root of (9.19) the right-hand side yields $\pm m(1 + x^2/y^2)$. The plus sign leads to EH waves and the minus to HE waves. In particular, for the HE_m modes the equation simplifies, for small σ, to

$$\frac{xJ_m(x)}{J_{m-1}(x)} = \frac{yK_m(y)}{K_{m-1}(y)}\left[1 + \frac{\sigma\epsilon_m}{2yK_m(y)K_{m-1}(y)}\right] \tag{9.21}$$

(Note that there is an error in Reference 5 of chapter 7 due to the omission of the factor 2 in the corresponding equation (52) *et seq*.)

In (9.21) the term in σ gives the effect of the fiber curvature.

9.4 Attenuation coefficient

Let x_0 and y_0 be the solution of (9.21) when $\sigma = 0$. Since we have, in general,

$$x^2 + y^2 = k_c^2 a^2 [(\epsilon/\epsilon_c) - 1] \tag{9.22}$$

then if (9.21) has a solution $x = x_0 + \Delta x$, $y = y_0 + \Delta y$, it follows from (9.22) that

$$x_0 \Delta x + y_0 \Delta y = 0 \tag{9.23}$$

to first order.

To the same order (9.21) gives

$$\Delta x \left[\frac{J_m}{J_{m-1}} + \frac{x_0 J'_m}{J_{m-1}} - \frac{x_0 J_m J'_{m-1}}{J^2_{m-1}}\right]$$
$$= \frac{\frac{1}{2}\epsilon_m \sigma}{K^2_{m-1}} + \Delta y \left[\frac{K_m}{K_{m-1}} + \frac{y_0 K'_m}{K_{m-1}} - \frac{y_0 K_m K'_{m-1}}{K^2_{m-1}}\right] \tag{9.24}$$

Putting Δy in terms of Δx via (9.23) and solving gives

$$\Delta x = \frac{\frac{1}{2}\epsilon_m \sigma x_0}{(x_0^2 + y_0^2)K_m K_{m-2}} \tag{9.25}$$

In terms of a radiative correction $\Delta k'$ to k' we get, since $k' \approx k_c \approx k(= \omega\sqrt{\epsilon\mu_0})$,

$$\frac{\Delta k'}{k'} = \frac{-\frac{1}{2}\sigma x_0^2 \epsilon_m}{(ka)^4 [(\epsilon/\epsilon_c) - 1] K_m K_{m-2}} \tag{9.26}$$

(Note that $K_{m-2} = K_{2-m}$ for $m < 2$). All terms in (9.26) are positive, so since x contains a factor j, (9.26) gives the attenuation coefficient directly as $-j\Delta k'$.

Since $\gamma^2 - 1 = (k'^2 - k^2)/k^2 = y_0^2/k^2a^2 \ll 1$, the expression for σ can be approximated by

$$\sigma = (j/4)\exp[-\tfrac{2}{3}kR_0(y_0/ka)^3]\left(\frac{\pi}{kR_0(y_0/ka)^3}\right)^{1/2} \tag{9.27}$$

For the HE_{11} mode, $m = 1$ and y_0 is small so that $K_1(y_0) \sim 1/y_0$. Then (9.26) and (9.27) give,

$$\frac{\Delta k'}{k'} \approx -j\exp[-\tfrac{2}{3}kR_0(y_0/ka)^3](\pi a/R_0)^{1/2}\frac{x_0^2 y_0^{1/2}}{4(ka)^3[(\epsilon/\epsilon_c) - 1]} \tag{9.28}$$

As an example of the rapid increase of attenuation per radian of arc, as the radius of curvature decreases, (9.28) gives, for the case $a = \lambda$, $\epsilon/\epsilon_c = 1{\cdot}02$,

$$R_0 = 2000\,\lambda,\ \text{attenuation} = 0{\cdot}005\ \text{neper/radian}$$

$$R_0 = 1000\,\lambda,\ \text{attenuation} = 0{\cdot}2\ \text{neper/radian.}$$

Results from (9.28) are independent of angle of the plane of the bend relative to the curvature. This result is strictly true only as $\epsilon_c \to \epsilon$; some additional comments are given at the end of Appendix F.

9.5 Curved dielectric guide of arbitrary cross-section

We have seen that the foregoing solution for the circular fiber depended not only on a knowledge of the fields inside and outside the core, but also on the construction of a secular (eigenvalue) equation from these fields which could be solved by a perturbation procedure. For a dielectric guide of noncircular cross-section and possibly inhomogeneous transverse permittivity distribution, the absence of separable field solutions and boundary conditions prevents us from applying a similar method in general. At present, such straight guides are usually analyzed numerically or (in some special cases) approximated analytically, and one has some knowledge of the fields and propagation constant for the straight waveguide. In the remainder of this chapter, curvature effects on surface waves which propagate along such guides will be determined entirely in terms of these known straight waveguide quantities.

The geometry is indicated in Fig. 7.6. In the core region ($r_{min} < r < r_{max}$) the dielectric permittivity $\epsilon = \epsilon_0 n^2$ is allowed to be a function of the coordinates r, z, while for $r < r_{min}$ and $r > r_{max}$ the permittivity is constant at $\epsilon_0 n_2^2$ and $\epsilon_0 n_0^2$ respectively. A global coordinate system $(r,\ \phi,\ z)$ whose origin coincides with the center of curvature of the guide, as well as two local ones (x, y, s) and $(\rho,\ \psi,\ s)$:

$$\left\{\begin{aligned} x &= -\rho \sin\psi = z \\ y &= \rho\cos\psi = r - R_0 \\ s &= R_0\phi \end{aligned}\right\}$$

is illustrated, with the radius of curvature R_0 defined as the distance between the origins of the global and local systems. For simplicity of interpretation the dielectric media will be assumed lossless so that any attenuation is totally due to radiation. In each co-ordinate system, the expression $\boldsymbol{a}_j$ is used to represent the unit vector in the j-direction. For instance, for the global co-ordinates, we have unit vectors $\boldsymbol{a}_r$, $\boldsymbol{a}_\phi$ and $\boldsymbol{a}_z$.

To start out, we recast Maxwell's equations for the curved guide into an integral form, similar to those from which perturbation formulas for closed waveguides are obtained.[3] The differential form of Maxwell's equations for an $\exp\,[j\omega t - j\bar{\gamma}k_0 R_0\phi]$ variation, is

$$\begin{aligned} \text{curl}_t \boldsymbol{E} - jk_0\bar{\gamma}\frac{R_0}{r}\,\boldsymbol{a}_\phi \times \boldsymbol{E} &= -j\omega\mu_0\boldsymbol{H} \\ \text{curl}_t \boldsymbol{H} - jk_0\bar{\gamma}\frac{R_0}{r}\,\boldsymbol{a}_\phi \times \boldsymbol{H} &= j\omega\mu\epsilon\boldsymbol{E} \end{aligned} \qquad (9.29)$$

where $k_0 = \omega\sqrt{\mu_0\epsilon_0}$, γ is the (as yet unknown) normalized complex propagation constant for the mode, and the subscript t denotes a transverse differential operator obtained by setting $\partial/\partial\phi$ equal to zero. Similarly, for the corresponding mode of the straight guide with an $\exp\,[j\omega t - jk_0\gamma_0 s]$ variation,

$$\begin{aligned} \nabla_t \times \boldsymbol{E}_0 - jk_0\gamma_0\boldsymbol{a}_s \times \boldsymbol{E}_0 &= -j\omega\mu_0\boldsymbol{H}_0 \\ \nabla_t \times \boldsymbol{H}_0 - jk_0\gamma_0\boldsymbol{a}_s \times \boldsymbol{H}_0 &= j\omega\epsilon\boldsymbol{E}_0 \end{aligned} \qquad (9.30)$$

where the subscript 0 on the fields and propagation constant indicates the straight guide, and the operator ∇_t operates on the local cartesian co-ordinates (with $\partial/\partial s = 0$) as if R_0 was infinite, i.e., as if (x, y, s) were a genuine rectangular co-ordinate system. The two rotation operators are related by

$$\text{curl}_t \boldsymbol{A} = \nabla_t \times \boldsymbol{A} + \frac{A_\phi}{r} \boldsymbol{a}_z \tag{9.31}$$

By symmetry, there also exist modes on the straight and curved guides travelling in the negative s (or ϕ) direction. On the straight guide, for example, we obtain this mode by taking $\gamma_0 \to -\gamma_0$, $\boldsymbol{E}_0 \to \boldsymbol{E}_0^-$ and $\boldsymbol{H}_0 \to \boldsymbol{H}_0^-$ in (9.30) where

$$\begin{aligned} \boldsymbol{a}_s \times \boldsymbol{E}_0^- &= \boldsymbol{a}_s \times \boldsymbol{E}_0 & \boldsymbol{a}_s \cdot \boldsymbol{E}_0^- &= -\boldsymbol{a}_s \cdot \boldsymbol{E}_0 \\ \boldsymbol{a}_s \times \boldsymbol{H}_0^- &= -\boldsymbol{a}_s \times \boldsymbol{H}_0 & \bar{\boldsymbol{a}}_s \cdot \boldsymbol{H}_0^- &= \boldsymbol{a}_s \cdot \boldsymbol{H}_0 \end{aligned} \tag{9.32}$$

and similarly on the curved guide, with s replaced by ϕ.

We now apply the angular two-dimensional divergence theorem (Appendix G) to the vector

$$\boldsymbol{F} = \boldsymbol{E}_0^- \times \boldsymbol{H} - \boldsymbol{E} \times \boldsymbol{H}_0^-$$

on an arbitrary surface S in a constant ϕ-plane; we obtain

$$\begin{aligned} \oint_C \frac{r}{R_0} [\boldsymbol{E}_0^- \times \boldsymbol{H} - \boldsymbol{E} \times \boldsymbol{H}_0^-] \cdot \boldsymbol{a}_n dl &= \int_S \frac{r}{R_0} \text{div}_t \boldsymbol{F} dS \\ = \int_S \left\{ jk_0 \left(\bar{\gamma} - \gamma_0 \frac{r}{R_0} \right) \boldsymbol{a}_s \cdot [\boldsymbol{E}_0 \times \boldsymbol{H} + \boldsymbol{E} \times \boldsymbol{H}_0] + \frac{1}{R_0} [E_z H_{0s} - E_{0s} H_z] \right\} dS \end{aligned} \tag{9.33}$$

where C is the boundary of S with outward unit normal $\boldsymbol{a}_n$, and (9.29) – (9.32) have been used. Furthermore, taking the scalar product of $\boldsymbol{H}$ and $\boldsymbol{E}$ with the first and second of (9.30) respectively and integrating over S, we have

$$\begin{aligned} -\frac{jk_0\gamma_0}{R_0} \int_S y \boldsymbol{a}_s \cdot [\boldsymbol{E}_0 \times \boldsymbol{H} + \boldsymbol{E} \times \boldsymbol{H}_0] \, dS \\ = -\frac{j}{R_0} \int_S y [\omega\mu_0 \boldsymbol{H}_0 \cdot \boldsymbol{H} + \omega\epsilon \boldsymbol{E}_0 \cdot \boldsymbol{E}] dS \\ + \frac{1}{R_0} \int_S y [\boldsymbol{E} \cdot \nabla_t \times \boldsymbol{H}_0 - \boldsymbol{H} \cdot \nabla_t \times \boldsymbol{E}_0] dS \end{aligned} \tag{9.34}$$

Finally, making use of the fact that

$$\nabla_t \cdot (y \boldsymbol{E}_0 \times \boldsymbol{H}) = \boldsymbol{a}_y \cdot \boldsymbol{E}_0 \times \boldsymbol{H} + y [\boldsymbol{H} \cdot \nabla_t \times \boldsymbol{E}_0 - \boldsymbol{E}_0 \cdot \nabla_t \times \boldsymbol{H}]$$

we can utilize (9.34) to express (9.33) in the form

$$-j\oint_C \left(1+\frac{y}{R_0}\right)[\bar{E_0}\times H - E\times \bar{H_0}]\cdot a_n dl$$

$$+\frac{1}{R_0}\int_S y[\omega\mu_0 H_0\cdot H + \omega\epsilon E_0\cdot E]dS$$

$$= k_0(\bar{\gamma}-\gamma_0)\int_S a_s\cdot[E_0\times H + E\times H_0]dS$$

$$+\frac{j}{R_0}\oint_C y[E_0\times H]\cdot a_n dl \tag{9.35}$$

$$-\frac{j}{R_0}\int_S y[E\cdot\nabla_t\times H_0 - E_0\cdot\nabla_t\times H]dS$$

$$-\frac{j}{R_0}\int_S [E_z H_{0s} - E_{0x}H_\phi]dS$$

9.6 Phase correction

As in the case of a slab guide, the first-order change in the real part of the propagation constant can be obtained by omitting terms that are exponentially small. This means we can replace E and H in (9.35) by E_0 and H_0 to obtain a first-order correction to γ_0:

$$k_0(\bar{\gamma}-\gamma_0) \cong \frac{\Delta}{P} \tag{9.36}$$

where

$$\Delta = \frac{1}{R_0}\int_S y[\omega\mu_0 H_0\cdot H_0 + \omega\epsilon E_0\cdot E_0]dS - \frac{j}{R_0}\oint_C y[E_0\times H_0]\cdot a_n dl \tag{9.37}$$

$$P = 2\int_S a_s\cdot E_0\times H_0 dS \tag{9.38}$$

We assume that S is chosen large enough so that it is permissible to approximately evaluate Δ and P by extending S to cover the entire transverse plane so that the contour integral in (9.37) vanishes by reason of the exponential decay of E_0 and H_0 and we have

$$\Delta = \frac{1}{R_0}\int y[\omega\mu_0 H_0\cdot H_0 + \omega\epsilon E_0\cdot E_0]\,dS \tag{9.39}$$

$$P = 2\int a_s\cdot E_0\times H_0 dS \tag{9.40}$$

where the absence of S on the integral signs denotes that the integrals are to be taken over the infinite transverse cross-sectional plane.

Appendix H confirms that Δ/P is purely real and is thus only a correction to the phase velocity and gives no information about radiation loss. The forms of (9.39) and (9.40) are suggestive, however, that Δ/P can be interpreted as a shift of the phase velocity reference point from $y_{ph} = 0$ to

$$y_{ph} = \frac{\int y[\omega\mu_0 \boldsymbol{H}_0 \cdot \boldsymbol{H}_0 + \omega\epsilon \boldsymbol{E}_0 \cdot \boldsymbol{E}_0]dS}{\int [\omega\mu_0 \boldsymbol{H}_0 \cdot \boldsymbol{H}_0 + \omega\epsilon \boldsymbol{E}_0 \cdot \boldsymbol{E}_0]dS} \tag{9.41}$$

the center of gravity of the energy-like quantity $[\mu_0 \boldsymbol{H}_0 \cdot \boldsymbol{H}_0 + \epsilon \boldsymbol{E}_0 \cdot \boldsymbol{E}_0]$. The amount of this shift depends, of course, on the choice of the point $y = 0$ (i.e. the choice of definition of the radius of curvature R_0). If the mode is symmetrical in y, then $\Delta = 0$ (Appendix H) and the phase correction is then at most of order $1/R_0^2$. We have seen this behaviour already for closed waveguides in Chapter 4. This interpretation of course agrees with the earlier observation concerning the shift in the propagation constant in the slab case (see section 8.5).

The phase correction can be important in several situations, when a quantity depending on the product $\bar{\gamma}R_0$ appears. For a finite section of bend, for example, we have seen in section 8.7 that an important factor in the scattering of a surface wave between two straight sections of guide is the excess phase shift on the curved section. It will be shown in the next section that the phase correction can have a marked effect on the attenuation as well.

9.7 Radiative correction

It has been shown that the radiation loss for the mode cannot be obtained through the use of $\boldsymbol{E}_0$ and $\boldsymbol{H}_0$ alone. Indeed, as in Chapter 4 the $1/R_0^2$, $1/R_0^3$, etc. corrections to γ_0 must all turn out to be real, unless the field which is reflected in quadrature from the caustic has been allowed for. To obtain the imaginary correction to γ_0, we return to (9.35), but instead of using just $\boldsymbol{E}_0$ and $\boldsymbol{H}_0$, we incorporate the local radiation condition into the equation by including the exponentially small fields produced by reflection of $\boldsymbol{E}_0$ and $\boldsymbol{H}_0$ from the caustic (see section 7.8). The spectrum functions of the reflected field components are given [to $0(1/k_0R_0)$] by (7.72):

$$\begin{aligned} \delta\tilde{E}_z(x,\xi) &= \mathcal{R}(\xi)e^{+\eta(\bar{\gamma},\xi)k_0 y}\,U(\xi)\tilde{E}(\xi) \\ \delta\tilde{H}_z(x,\xi) &= \mathcal{R}(\xi)e^{+\eta(\bar{\gamma},\xi)k_0 y}\,U(\xi)\tilde{H}(\xi) \end{aligned} \qquad (y > y_{max}) \tag{9.42}$$

where $\mathcal{R}(\xi)$ is given by (7.66), $\eta(\bar{\gamma}, \xi)$ is given by (7.63), $\tilde{E}(\xi)$ and $\tilde{H}(\xi)$ are shortened notations for $\tilde{E}_{z0}$ and $\tilde{H}_{z0}$ respectively, and $U(\xi)$ is a step function which is 1 for $|\xi| < 1$ and 0, otherwise.* We saw in section 7.6 that no corresponding term in the spectrum function is introduced on the inner side of the bend (except as required to assure continuity with the fields on the outer side), so that δE and δH are zero for $y < y_{min}$. The remaining field components are related to δE_z and δH_z by (7.71).

The spectral representation for the fields is taken advantage of by choosing the surface S to be the strip $-\infty < z < \infty$, $R_0 - y_0 < r < R_0 + y_0$, with boundary $C_1 + C_2$. The quantity $y_0 > \max\{|y_{max}|, |y_{min}|\}$ is some fixed number chosen to satisfy the following conditions:

(1) $|k_0 y_0 \sigma_0| \ll 1$ (where σ_0 is given by (7.68)), so that δE and δH when included do not change the values of (9.37) and (9.38),

(2) $|y_0/R_0| \ll 1$, so that higher order terms in (y_0/R_0) may be neglected,

(3) $|e^{-2k_0\eta(\bar{\gamma},0)y_0}/\eta(\bar{\gamma}, 0)| \ll 1$, in order to be able to neglect the contour integral in (9.37) so that (9.39) will hold.

It is easily seen that the condition

$$|2\tau| \gg 1 \tag{9.43}$$

where τ is the exponent appearing in $\mathcal{R}(\xi)$ given by (7.68), is sufficient to insure that (1)–(3) can be satisfied; in the case when $|\eta(\bar{\gamma}, 0)| \ll 1$, for example, which is typical of optical waveguides whose core and cladding refractive indices are very close, the choice $k_0 y_0 = |1/\eta^2(\bar{\gamma}, 0)|$ will do.

Substiuting $\boldsymbol{E}_0 + \delta \boldsymbol{E}$ and $\boldsymbol{H}_0 + \delta \boldsymbol{H}$ into (9.35) subject to the above conditions results in

$$k_0(\bar{\gamma} - \gamma_0) \simeq \frac{\Delta}{P} - j\frac{c}{P} \tag{9.44}$$

where Δ and P are still given by (9.39) and (9.40), while

$$c = \int_{-\infty}^{\infty} [\boldsymbol{E}_0^- \times \delta \boldsymbol{H} - \delta \boldsymbol{E} \times \boldsymbol{H}_0^-] dx_{y=y_0} \tag{9.45}$$

In deriving (9.45), terms involving $\delta \boldsymbol{E}/R_0$ and $\delta \boldsymbol{H}/R_0$ are omitted. We also note that the integration in (9.45) is defined only along the path $y = y_0$ exterior to the curved guide. Because of the absence of caustics on the inner side of the bend, no contribution to c comes from the inner portion C_2 of the contour.

* As in Chapter 7, the tilde is used to denote the Fourier transform with respect to z

Expression (9.45) can be transformed using the convolution theorem for the Fourier transform pair (7.59):

$$\int_{-\infty}^{\infty} A(z)B(z)dz = \frac{2\pi}{k}\int_{-\infty}^{\infty} \widetilde{A}(\xi)\widetilde{B}(-\xi)d\xi = \frac{2\pi}{k}\int_{-\infty}^{\infty} \widetilde{A}(-\xi)\widetilde{B}(\xi)d\xi \tag{9.46}$$

for any suitable functions A and B and their transforms $\widetilde{A}$ and $\widetilde{B}$. Making use of (7.71) to obtain the components E_ϕ and H_ϕ, c is then expressed as a spectral integral:

$$c = \frac{4\pi j}{k_0 n_0^3}\int_{-1}^{1} \frac{\eta(\gamma_0, \xi)\mathcal{R}(\xi)}{(1-\xi^2)}\left[n_0^2\sqrt{\frac{\epsilon_0}{\mu_0}}\,\widetilde{E}(\xi)\widetilde{E}(-\xi) + \sqrt{\frac{\mu_0}{\epsilon_0}}\,\widetilde{H}(\xi)\widetilde{H}(-\xi)\right]d\xi \tag{9.47}$$

The locally growing contributions (9.42) have multiplied the locally decaying portions of the unperturbed field in (9.45) and within assumptions (1)–(3) above, (9.47) holds independently of the choice of γ_0. The integral in (9.47) is from -1 to 1 due to the step function $U(\xi)$ in the spectrum representation.

As evident from (7.66) of Chapter 7, the strong exponential dependence of $\mathcal{R}(\xi)$ suggests that the integral in (9.47) may be evaluated approximately by the saddle-point method, provided $\widetilde{E}$ and $\widetilde{H}$ are slowly varying functions of ξ. For this latter statement to be true, E_x and H_x must be sufficiently rapidly decaying in the x-direction, by virtue of the known properties of Fourier transforms. In a slab waveguide, for example, whose fields are independent of x, the Fourier transforms are Dirac δ-functions, and the saddle-point technique will be inapplicable. We thus restrict ourselves to consideration of modes whose characteristic dimension in the x-direction (and for low-order modes this is roughly the dimension of the guide itself) is small compared to the radius of curvature R_0. Slab waveguides, of course, can be treated without the need for spectrum functions or saddle-point integrations, in a much simpler manner.

Equation (9.47) can be rewritten using (7.66) as

$$c = \frac{2\pi\omega}{k_0^2 n_0^3}\int_{-1}^{1}[\epsilon_0 n_0^2\,\widetilde{E}(\xi)\widetilde{E}(-\xi) + \mu_0\widetilde{H}(\xi)\widetilde{H}(-\xi)]\,\frac{\eta(\gamma, \xi)\exp[k_0 R_0 q(\xi)]}{1-\xi^2}\,d\xi \tag{9.48}$$

where

$$q(\xi) = 2\left[\eta(\bar{\gamma}, \xi) + \frac{\bar{\gamma}}{2}\log\frac{\bar{\gamma}-\eta(\bar{\gamma}, \xi)}{\bar{\gamma}+\eta(\bar{\gamma}, \xi)}\right] = 2\left[\eta(\bar{\gamma}, \xi) + \bar{\gamma}\tanh^{-1}\left(\frac{\eta(\bar{\gamma}, \xi)}{\bar{\gamma}}\right)\right] \tag{9.49}$$

As in section 7.8, it is easily verified that the steepest-descent path in the ξ-plane lies very close to the real axis between $\xi = \pm 1$ (precisely the real axis if $\bar{\gamma}$ were purely real) and that $q(\xi) \to -\infty$ as $\xi \to \pm 1$. The saddle point (at which $q'(\xi) = 0$) is located at $\xi = 0$; two spurious zeroes of $q'(\xi)$ appear near the imaginary axis where $\eta(\bar{\gamma}, \xi) = 0$. These latter must be sufficiently far away from $\xi = 0$ for the saddle-point method to apply, and a closer investigation shows that condition (9.43) suffices to insure this. The result of the saddle-point integration is

$$c = \left(\frac{\pi\eta_0}{k_0 R_0}\right)^{1/2} \frac{2\pi\omega}{k_0^2 n_0^3}\{\epsilon_0 n_0^2 [\tilde{E}(0)]^2 + \mu_0 [\tilde{H}(0)]^2\}\exp[-2\tau] \tag{9.50}$$

where

$$\tau = -k_0 R_0\left\{\eta + \frac{\bar{\gamma}}{2}\log\frac{\bar{\gamma}-\eta}{\bar{\gamma}+\eta}\right\}$$

$$= k_0 R_0 \bar{\gamma}\left\{\frac{1}{3}\left(\frac{\eta}{\bar{\gamma}}\right)^3 + \frac{1}{5}\left(\frac{\eta}{\bar{\gamma}}\right)^5 + \frac{1}{7}\left(\frac{\eta}{\bar{\gamma}}\right)^7 + \ldots\right\} \tag{9.51}$$

and we have abbreviated $\eta = \eta(\bar{\gamma}, 0)$ and $\eta_0 = \eta(\gamma_0, 0)$. Note that it is *not* in general permissible to retain only the first term of this series expansion, even if $|\eta/\bar{\gamma}| \ll 1$, because for large k_0R_0 the effect of neglecting $k_0R_0(\eta/\bar{\gamma})^5$ in the exponent of (9.50) could very well be important.

Equations (9.44), (9.39), (9.40), (9.50) and (9.51) now give the general solution for the phase correction and radiative attenuation due to curvature of the waveguide, requiring knowledge only of the fields $\boldsymbol{E}_0$, $\boldsymbol{H}_0$ and propagation constant γ_0 of the straight guide. It would ordinarily suffice to evaluate τ in (9.51) by replacing $\bar{\gamma}$ by γ_0. However, for a nonsymmetric mode ($\Delta \neq 0$), we have by (9.44) that $\bar{\gamma} - \gamma_0$ is $0(1/k_0R_0)$, and this difference will result in an additional term of $0(1)$ in the exponent of (9.50). Hence in this case, we obtain τ by a Taylor series expansion of (9.51) about γ_0.

$$\tau \simeq \tau_0 - \frac{1}{2}k_0R_0(\bar{\gamma}-\gamma_0)\log\frac{\gamma_0-\eta_0}{\gamma_0+\eta_0} \simeq \tau_0 - \frac{1}{2}R_0\left(\frac{\Delta}{P}\right)\log\frac{\gamma_0-\eta_0}{\gamma_0+\eta_0}$$

$$\simeq \tau_0 + R_0\left(\frac{\Delta}{P}\right)\left[\frac{\eta_0}{\gamma_0} + \frac{1}{3}\left(\frac{\eta_0}{\gamma_0}\right)^3 + \frac{5}{1}\left(\frac{\eta_0}{\gamma_0}\right)^5 + \ldots\right] \tag{9.52}$$

and $\tau_0 = \tau(\bar{\gamma} = \gamma_0)$. If $|\eta_0/\gamma_0| \ll 1$, the retention of the first term only in the series expansion is in this case legitimate.

Equation (9.50) requires that we have accurate knowledge of the fields external to the guide, whereas (9.39) and (9.40) are, for modes not close to cutoff, primarily determined by the fields inside the guide. In many cases (for example, the rectangular dielectric channel) there are no closed form analytic expressions available for the fields, and either approximate analytic forms or numerical solutions are needed. In either event, the internal guide fields will generally be known more accurately than the external fields, and expressions for $\tilde{E}(0)$ and $\tilde{H}(0)$ in terms of the internal fields should provide more accurate results in these cases. These can be obtained for the case when the cladding is homogeneous ($n_2 = n_0$) by using integral representations for the modal fields due to Katsenelenbaum:[4]

$$E_0(\boldsymbol{\rho}) = -\frac{j}{4}\,[\text{grad div} + k_0^2 n_0^2] \int_S \left(\frac{n^2(\rho')}{n_0^2} - 1\right) E_0(\boldsymbol{\rho}') H_0^{(2)}[-jk_0\eta_0 |\boldsymbol{\rho} - \boldsymbol{\rho}'|]\,dS'$$

$$H_0(\boldsymbol{\rho}) = \frac{\omega\epsilon_0 n_0^2}{4}\,\text{curl} \int_S \left(\frac{n^2(\rho')}{n_0^2} - 1\right) E_0(\boldsymbol{\rho}') H_0^{(2)}[-jk_0\eta_0 |\boldsymbol{\rho} - \boldsymbol{\rho}'|]\,dS' \tag{9.53}$$

where $\boldsymbol{\rho}$ and $\boldsymbol{\rho}'$ are observation and integration points in the transverse plane, the differential operators act on $\boldsymbol{\rho}$ but not $\boldsymbol{\rho}'$ and $\partial/\partial s$ is to be understood as replaced everywhere by multiplication by $-jk_0\gamma_0$. The field $E_0(\boldsymbol{\rho}')$ need only be known over the core region, since $n^2(\boldsymbol{\rho}') = n_0^2$ outside this region.

Utilizing the plane wave spectrum representation for $H_0^{(2)}$:

$$H_0^{(2)}[-jk_0\eta_0 \{(x - x')^2 + (y - y')^2\}^{1/2}] = -\frac{n_0}{j\pi} \int_{-\infty}^{\infty} e^{-jk\xi(x - x') - k_0\eta(\gamma_0, \xi)|y - y'|} \frac{d\xi}{\eta(\gamma_0, \xi)} \tag{9.54}$$

the differentiation $\partial/\partial y$ becomes $-k_0\eta(\gamma_0, \xi)$ for $y > y'$ and $\partial/\partial x$ becomes $-jk\xi$, so that the transformed fields can be identified from (9.53) as

$$\tilde{E}(\xi) = \frac{k_0^2}{4\pi\eta(\gamma_0, \xi)} \int_S [n^2(\boldsymbol{\rho}') - n_0^2]\{j\xi\eta(\gamma_0, \xi)E_{0y}(\boldsymbol{\rho}') - \xi\gamma_0 E_{0s}(\boldsymbol{\rho}') + n_0(1 - \xi^2)E_{0x}(\boldsymbol{\rho}')\}\, e^{jk\xi x' + k_0\eta(\gamma_0, \xi)y'}\, dS' \tag{9.55}$$

$$\tilde{H}(\xi) = \frac{jk_0^2 n_0}{4\pi\eta(\gamma_0, \xi)}\sqrt{\frac{\epsilon_0}{\mu_0}}\int_S [n^2(\boldsymbol{\rho}') - n_0^2]\{-\eta(\gamma_0, \xi)E_{0s}(\boldsymbol{\rho}')$$
$$+ j\gamma_0 E_{0y}(\boldsymbol{\rho}')\}e^{jk\xi x' + k_0\eta(\gamma_0,\xi)y'}dS'$$

In particular, we have the relatively simple formulae

$$\tilde{E}(0) = \frac{k_0^2 n_0}{4\pi\eta_0}\int_S [n^2(\boldsymbol{\rho}') - n_0^2]E_{0x}(\boldsymbol{\rho}')e^{k_0\eta_0 y'}dS' \tag{9.56}$$

$$\tilde{H}(0) = -\frac{k_0^2 n_0}{4\pi\eta_0}\sqrt{\frac{\epsilon_0}{\mu_0}}\int_S [n^2(\boldsymbol{\rho}') - n_0^2]\{\gamma_0 E_{0y}(\boldsymbol{\rho}')$$
$$+ j\eta_0 E_{0s}(\boldsymbol{\rho}')\}e^{k_0\eta_0 y'}dS' \tag{9.57}$$

When the general formulas of this section are used for the circular fiber, substantially the same expression for the attenuation coefficient is obtained as in section (9.4). In the weakly guiding limit (when core and cladding indices become equal) the attenuation is predicted to be polarization independent, but the breaking of the polarization degeneracy described in Appendix F cannot be accounted for by the present method.

9.8 Rectangular dielectric channel

A more instructive application of the formulas of the preceding sections is to the rectangular fiber, for which only approximate analytical forms for the fields are available. These forms can be found in the book by Marcuse[5] and will not be repeated here, but only the results of their substitution into the previous formulae will be given.

For a rectangular guide whose x-dimension is $2a$, whose y-dimension is $2b$, with core index n_1 surrounded by a medium of index n_0, the propagation constant γ_0 is given approximately by the solution of the following equations if the mode is not too near cut-off:

$$E_{pq}^x: \tan 2\kappa_x A = \frac{2n_0^2 n_1^2 \kappa_x \eta_x}{n_0^4\kappa_x^2 - n_1^4\eta_x^2}$$
$$\tan 2\kappa_y B = \frac{2\kappa_y\eta_y}{\kappa_y^2 - \eta_y^2} \tag{9.58}$$

$$E_{pq}^y: \tan 2\kappa_x A = \frac{2\kappa_x\eta_x}{\kappa_x^2 - \eta_x^2} \tag{9.59}$$

$$\tan 2\kappa_y B = \frac{2n_0^2 n_1^2 \kappa_y \eta_y}{n_0^4 \kappa_y^2 - n_1^4 \eta_y^2}$$

where $A = k_0 a$, $B = k_0 b$ and

$$\begin{aligned} \eta_x^2 + \kappa_x^2 &= \eta_y^2 + \kappa_y^2 = n_1^2 - n_0^2 \\ \gamma_0^2 &= n_1^2 - \kappa_x^2 - \kappa_y^2 \end{aligned} \tag{9.60}$$

The mode designations E_{pq}^x and E_{pq}^y refer to the direction (x or y) in which the E-field is primarily polarized, while the mode indices p and q refer to the number of maxima of the dominant E-field in the x and y directions respectively.

The resulting attenuation coefficient α as obtained from $\alpha = |c/P|$ for the E_{pq}^x modes is

$$\alpha = n_0 \left(\frac{1}{\pi k_0 R_0}\right)^{1/2} e^{-2\tau_0} \left(\frac{\kappa_y^2}{\kappa_x^2}\right) \frac{4n_1^2(n_0^2 + \eta_x^2)F^x}{k_0^2 A_e^x \gamma_0 \eta_0^{3/2} \eta_x^2 [n_0^4 + \eta_x^2(n_1^2 + n_0^2)]} \tag{9.61}$$

where the effective area A_e^x is given by

$$\begin{aligned} A_e^x = 4\Bigg\{ & ab + \frac{a}{k_0 \eta_y}\left[1 - \frac{\kappa_x^2 \kappa_y^2}{n_0^2(n_0^2 + \eta_x^2)}\right] \\ & + \frac{b}{k_0 \eta_x}\left[1 + \frac{\kappa_x^2(n_0^2 + n_1^2) - n_1^2(n_1^2 - n_0^2)}{n_0^4 + \eta_x^2(n_0^2 + n_1^2)}\right] + \frac{1}{k_0^2 \eta_x \eta_y} \\ & \left[1 + \frac{n_0^2(n_1^2 - n_0^2) - \eta_x^2(n_0^2 + n_1^2) - \left(\dfrac{\kappa_x^2}{n_1^2 - n_0^2}\right)\left(\dfrac{\kappa_y^2}{n_1^2 - n_0^2}\right)\left(n_0^2 n_1^2 + \dfrac{n_1^2 \eta_x^2 (n_1^2 - n_0^2)}{n_0^2 + \eta_x^2}\right)}{n_0^4 + \eta_x^2(n_0^2 + n_1^2)}\right]\Bigg\} \end{aligned} \tag{9.62}$$

$$\text{and} \quad F^x = \begin{cases} 0 & p \text{ even} \\ [\eta_0 \sinh \eta_0 B + \eta_y \cosh \eta_0 B]^2 & p \text{ odd}, q \text{ odd} \\ [\eta_0 \cosh \eta_0 B + \eta_y \sinh \eta_0 B]^2 & p \text{ odd}, q \text{ even} \end{cases} \tag{9.63}$$

Similarly, the attenuation coefficient for the E_{pq}^y modes is

$$\alpha = n_0 \left(\frac{1}{\pi k_0 R_0}\right)^{1/2} e^{-2\tau_0} \left(\frac{\kappa_y^2}{\kappa_x^2}\right) \frac{4\gamma_0^2(n_0^2 + \eta_y^2)F^y}{k_0^2 A_e^y \gamma_0 \eta_0^{3/2} \eta_x^2 [n_0^4 + \eta_y^2(n_0^2 + n_1^2)]} \tag{9.64}$$

where A_e^y is obtained from (9.62) by interchanging $a \leftrightarrow b$ and $x \leftrightarrow y$, and

$$F^y = \begin{cases} 0 & p \text{ even} \\ \left[\eta_0 \dfrac{n_0}{n_1} \sinh \eta_0 B + \eta_y \dfrac{n_1}{n_0} \cosh \eta_0 B\right]^2 & p \text{ odd}, q \text{ odd} \\ \left[\eta_0 \dfrac{n_0}{n_1} \cosh \eta_0 B + \eta_y \dfrac{n_1}{n_0} \sinh \eta_0 B\right]^2 & p \text{ odd}, q \text{ even} \end{cases} \tag{9.65}$$

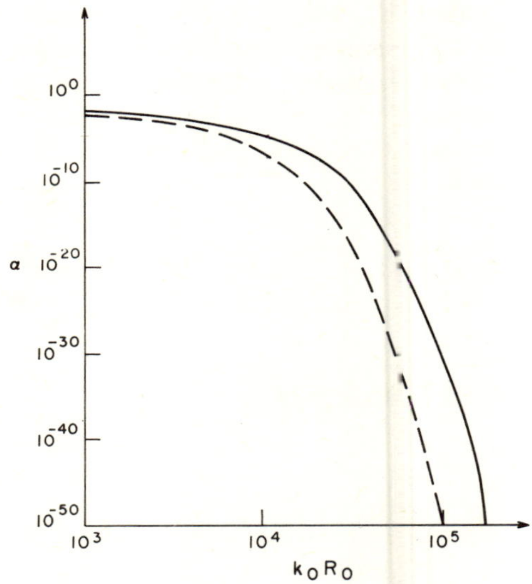

Fig. 9.1 Attenuation coefficient for E_{11} mode of square dielectric channel ($k_0a = k_0b = 14{\cdot}1$; $n_1 = 1{\cdot}01$; $n_0 = 1{\cdot}00$,) —— eqn. (9.61); ---- Marcatili's formula

to get to α in nepers, multiply by 2k₀

Formulas (9.61) and (9.64) bear a striking resemblance to that for the circular fiber, (9.26): each carries the dominating exponential dependence on R_0, a further factor of $R_0^{-1/2}$, an effective area, and an exponential dependence upon the thickness of the guide in the plane of the bend. The apparent vanishing of α for modes with p even is probably due to the fact that approximate, LP (linearly-polarized) field forms have been used. A similar phenomenon occurs for the circular fiber when LP modes are used,[6] but exact fiber modes are recovered by superposing two such modes in a specific ratio, and the exact modes, as shown in section 9.4, are attenuated in a virtually polarization-independent manner. It seems reasonable to conjecture that a similar thing happens for the rectangular guide, and for an exact mode one may have to take an average of F^x or F^y between the values for p odd and p even.

These formulas are valid if the guide cross-section is not so large that more than a few guided modes are supported (since otherwise severe distortion of the internal fields begins to occur). As mentioned in the previous section, they break down as $2a \to \infty$ [since $\kappa_x \to 0$ and $\kappa_x A \to$ constant by (9.58) or (9.59)] and in this limit the loss more closely resembles that of a slab, and the formulae of Marcatili[7] apply. In Fig. 9.1, we give a comparison between (9.61) and the formula of Marcatili for an E^x_{11} mode of a square dielectric channel ($k_0 a = k_0 b = 14{\cdot}1$; $n_1 = 1{\cdot}01$, $n_0 = 1{\cdot}00$). The major difference between the curves arises because of the factor of roughly $\exp(-\frac{2}{3}\eta_0^3 k_0 R_0)$ appearing in (9.61) as compared with $\exp(-\frac{2}{3}\eta_y^3 k_0 R_0)$ in Marcatili's, where η_y is the slab decay coefficient for fields in the y direction in his approximation. Even for moderate values of $k_0 R_0$, the discrepancy is substantial enough to be important for design purposes.

References

1 Kao, K.C., and Hockham, G.A.: 'Dielectric-fiber surface waveguides for optical frequencies', *Proc. IEE*, Vol. 113, 1966, pp. 1151–1158

2 Clarricoats P.J.B.: 'Propagation along unbounded and bounded dielectric rods', *Proc. IEE*, Vol. 108C, 1960, pp. 170–176

3 Harrington, R.F.: 'Time harmonic electromagnetic fields' (McGraw Hill, New York, 1961), pp. 326–345

4 Katsenelenbaum, B.Z.: 'On the propagation of electromagnetic waves along an infinite dielectric cylinder at low frequencies', *Dokl. Akad. Nauk SSSR*, Vol. 58, 1947, pp. 1317–1320

5 Marcuse, D.: 'Theory of dielectric optical waveguides' (Academic Press, New York, 1974), pp. 48–59

6 Kuester, E.F., Chang, D.C.: 'Surface-wave radiation loss from curved dielectric slabs and fibers,' *IEEE J. Quantum Electron.*, Vol. 11, 1975, pp. 903–907

7 Marcatili, E.A.J.: 'Bends in optical dielectric guides,' *Bell Syst. Tech. J.*, Vol. 48, 1969, pp. 2103–2132

Chapter 10

Modal field representations on curved open structures

10.1 Exact modes of a straight reactive surface

The last three chapters have been concerned with the changes produced by curvature of the waveguide axis upon a guided surface mode on an open waveguide, somewhat analogous to the study of closed waveguide modes done in Chapters 4 and 6. When, for a closed waveguide, we considered the junction between straight and curved guides in Chapter 5, it was found that a complete spectrum of modes for the curved guide was needed in addition to the (incident) mode of interest in order to describe the scattering at the junction. Because the essential character of these modes is not altered by a bend of finite radius, a standard (i.e., mode-matching) approach suffices to determine this scattering.

For open waveguides the situation is different in that the natural modes of a straight guide consist of a (usually) finite number of discrete surface modes, and an additional continuum of radiation modes whose wavenumbers lie on an infinite line in the complex plane.[1] We shall see in this section that the character of the modes on curved open structures is very different from those of straight structures, and that some modifications of the technique of Chapter 5 will be needed to treat curvature discontinuities in open waveguides.

Much of the investigations contained in this chapter as well as the next is comparatively recent, and because of the complexity of the problems involved, only very simple geometries will be considered here. Moreover, and especially in Chapter 11, more than the usual amount of detail will be omitted in view of the tentative nature of some aspects of the technique. There will also be changes in notation in these chapters as compared with previous ones: ν will now be used in place of γ to denote a normalized propagation constant; γ_0 will now be simply a normalized surface reactance. The radius of curvature will be R (not R_0), and the co-ordinate systems will be those indicated in Fig. 10.1.

In accordance with these comments, we will limit ourselves in the next two chapters to consideration of what is perhaps the simplest of surface waveguides, the reactive surface. TE waves satisfy the boundary condition (we consider the two-dimensional problem with $\partial/\partial z \equiv 0$)

$$\left(\frac{\partial}{\partial n} + k\gamma_0\right)E_z\bigg|_{n=0} = 0 \tag{10.1}$$

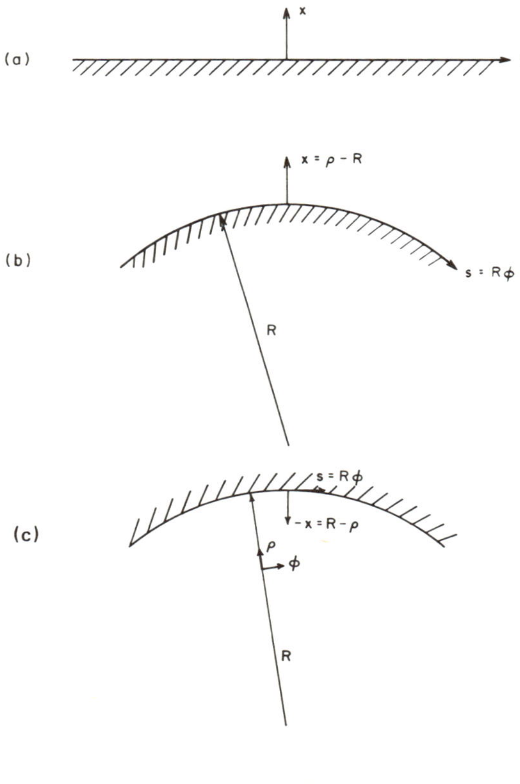

Fig. 10.1 (*a*) Infinitely extended straight impedance surface
(*b*) Infinitely extended curved impedance surface (convex)
(*c*) Infinitely extended curved impedance surface (concave)

at such a surface, where n is the outward normal distance from the surface, and γ_0 is a dimensionless positive quantity related to the surface reactance (the details of this are of no consequence for the present discussion, whose purpose is to illustrate some basic principles). We may use (10.1) with either the straight or the curved surfaces shown in Fig. 10.1.

Let us first recall the modal structure of the straight surface. The Green's function $G_1(\boldsymbol{\rho}; \boldsymbol{\rho}')$ satisfying

$$(\nabla^2 + k^2)\, G_1(\boldsymbol{\rho}; \boldsymbol{\rho}') = -\delta(\boldsymbol{\rho} - \boldsymbol{\rho}') \tag{10.2}$$

along with condition (10.1) and the radiation condition as $|\boldsymbol{\rho}| \to \infty$ (with $\boldsymbol{\rho} = x\mathbf{a}_x + y\mathbf{a}_y$) can be constructed as a Fourier integral with respect to the propagation direction (y):

$$G_1 = -\frac{j}{4\pi} \int_{C_1} e^{-jk\nu|y-y'|-jkwx_>} \{e^{jkwx_<} + \Gamma_0(\nu) e^{-jkwx_<}\} \frac{d\nu}{w}$$

$$w = (1-\nu^2)^{1/2}; \quad \mathrm{Im}(w) \leqslant 0; \quad x_{\gtrless} = \frac{\max}{\min}\{x, x'\}; \tag{10.3}$$

$$\Gamma_0(\nu) = \frac{w - j\gamma_0}{w + j\gamma_0} \tag{10.4}$$

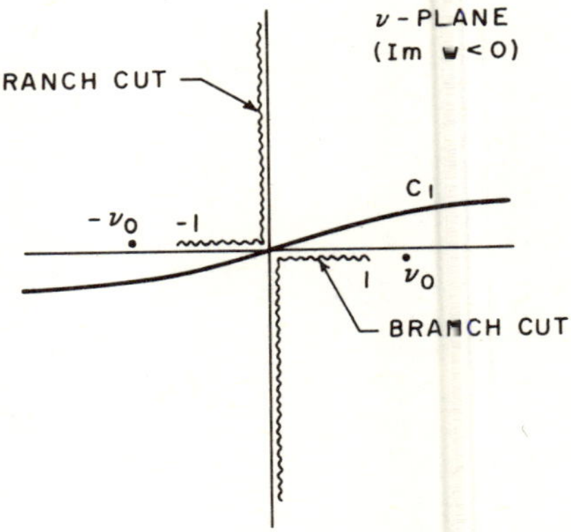

Fig. 10.2 Integration contour and singularities of integrand for G_1

The disposition of the integration path C_1 with respect to the singularities of the integrand is shown in Fig. 10.2. The choice of branch cuts is motivated by a desire to be able to evaluate (10.3) by deforming C_1 into the lower half-plane, for *any* $\boldsymbol{\rho}$ and $\boldsymbol{\rho}'$ (except, of course, for $\boldsymbol{\rho} = \boldsymbol{\rho}'$, where G_1 is infinite). The poles at $\pm\nu_0 = \pm(1+\gamma_0^2)^{1/2}$ occur as zeroes of the denominator of Γ_0, and upon deforming C_1, the residue contribution at ν_0 takes the form of a surface wave propagating along the surface in the y-direction.

If the contour deformation is carried out, and a change of integration variable performed on the branch cut integral, we obtain

$$G_1 = -j\frac{\gamma_0}{\nu_0}\Psi_{s0}(x)\,\Psi_{s0}(x')\,e^{-jk\nu_0|y-y'|}$$

$$-\frac{j}{2}\int_0^\infty \Psi_{c0}(w,x)\,\Psi_{c0}(w,x')\,e^{-jk\nu|y-y'|}\frac{dw}{\nu N^2(w)} \quad (10.5)$$

where
$$N^2(w) = \pi(w^2+\gamma_0^2)/2$$
$$\nu = (1-w^2)^{1/2} \qquad 0<w<1$$
$$= -j(w^2-1)^{1/2} \qquad 1<w<\infty$$

and we can identify the representation (10.5) as a generalization of the well-known 'sum of products of mode functions' form of the Green's function for closed waveguides. The surface wave mode

$$\Psi_{s0}(x) = \exp(-k\gamma_0 x) \quad (10.6)$$

enters in the usual way, while for the continuous or radiation modes

$$\Psi_{c0}(w,x) = \tfrac{1}{2}[(w+j\gamma_0)\exp(jkwx)+(w-j\gamma_0)\exp(-jkwx)] \quad (10.7)$$

and the summation goes over into an integral.

Although derived from the line source excitation of a surface of infinite extent, by superposition these modes will describe the fields of fairly arbitrary source distributions, and will also apply to finite sections of surface if equivalent sources at the ends (aperture fields) can be set up. In this way, as for ordinary closed waveguide problems, the junction between two different waveguides can be treated within the modal framework. This is especially useful because of the following orthogonality properties which are easily verified:

$$\int_0^\infty \Psi_{s0}^2(x)\,dx = 1/(2k\gamma_0) \quad (10.8)$$

$$\int_0^\infty \Psi_{c0}(w,x)\,\Psi_{s0}(x)\,dx = 0 \quad (10.9)$$

$$\int_0^\infty \Psi_{c0}(w,x)\,\Psi_{c0}(w',x)\,dx = N^2(w)\delta(w-w')/k \quad (10.10)$$

where $\delta(w-w')$ is the Dirac delta-function. A wide variety of transitions can be treated using these straight waveguide modes.[1]

10.2 Exact modes of a convex reactive surface

If we formally take a convex reactive surface as in Fig. 10.1*b* which is allowed to be of infinite extent ($-\infty<\phi<\infty$), the Green's function

G_2 for this case can be constructed as a Fourier integral (now with respect to ϕ) as in the previous section. Although the intent of the infinitely extended surface is the same as before (to exclude the possibility of reflected waves coming from somewhere down the guide), the result in this case is a nonphysical situation, a kind of Riemann surface in which the point $(\rho, \phi + 2\pi)$ is not the same as (ρ, ϕ), and periodicity of the fields is dropped in favor of a 'radiation condition' as $\phi \to \pm\infty$. If it is preferred to deal only in a physical space, a set of perfect absorbers may be thought of at $\phi = \pm\pi$, completely eliminating any reflected fields or fields which have travelled one or more times around the cylinder.[2]

The Fourier representation is carried out now to obtain a formula[3] analogous to (10.3):

$$G_2 = -\frac{jkR}{8} \int_{C_2} e^{-jk\nu|s-s'|} H^{(2)}_{\nu kR}(k\rho_>) \{H^{(1)}_{\nu kR}(k\rho_<) - H^{(2)}_{\nu kR}(k\rho_<) P(\nu, R)\} d\nu$$

$$s = R\phi; \rho_{\gtrless} = \begin{matrix}\max\\ \min\end{matrix} \{\rho, \rho'\}; \tag{10.11}$$

$$P(\nu, R) = \frac{H^{(1)'}_{\nu kR}(kR) + \gamma_0 H^{(1)}_{\nu kR}(kR)}{H^{(2)'}_{\nu kR}(kR) + \gamma_0 H^{(2)}_{\nu kR}(kR)} \tag{10.12}$$

Here, R is the radius of curvature of the reactive surface, as indicated in Fig. 10.1*b*. The integration contour C_2 is shown in Fig. 10.3.

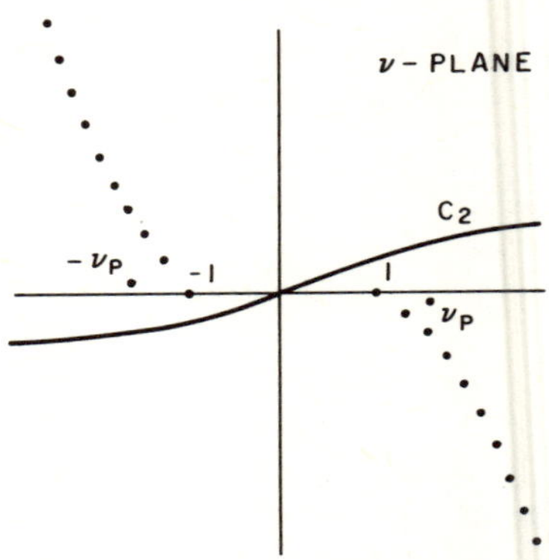

Fig. 10.3 Integration contour and singularities of integrand for G_2

The major difference between the integrands of (10.3) and (10.11) is that *the latter has no branch points in the ν-plane, but rather an infinite set of poles.* The absence of branch points follows because the

Hankel functions are entire functions of their index. On the other hand, the poles, whose positions[3,4] are roughly sketched in Fig. 10.3 and which must now serve to describe an arbitrary field over the convex surface, represent anything but a small perturbation of the radiation modes of the straight surface, whose place they must take. Partly as a result of this, these 'creeping-wave' modes are very slowly convergent when used to expand, say, the primary field of a point source in a geometrically illuminated region. The only exception to this is the pair of roots $\pm \nu_p$ which are indeed small perturbations to $\pm \nu_0$, and in fact are the weakly radiating surface waves which have been studied Chapters 7–9. In order to conveniently obtain a solution to the curvature discontinuity problem, which will be tackled in the next chapter, it would be desirable to have a representation for the fields of the convex section which is similar to (10.5) for the straight section, making a perturbation solution for large radius of curvature possible using the technique of Chapter 5.

On purely physical or intutitve grounds, we should expect that such a representation must exist which is valid at least in a 'boundary layer'[5] near the surface, where the curved surface 'looks' almost flat to both source and observer. In fact, if one *assumes* a representation analogous to (4.3), it is possible to formally derive a series for a 'radiation mode' in inverse powers of R; however (and this applies to the formal series used in Chapters 4–6 as well) it would be desirable to know when, if at all, such series are valid. For this reason sections 10.4 and 10.5 will be devoted to deriving an approximate 'mode' set valid in a boundary layer for describing fields on a convex reactive surface.

10.3 Exact modes of a concave reactive surface

The mode structure (singularities of the Green's function) for a concave reactive surface may be found in much the same way as in the previous section again using a Fourier representation with respect to the angle ϕ, which varies from $-\infty$ to $+\infty$ as before. This situation, illustrated in Fig. 10.1*c*, has the slight additional complication of a singularity in the space at the origin ($\rho = 0$). For many purposes, especially when the radius of curvature R of the surface is large, the only important region in scattering and waveguiding problems is $0 \leqslant \rho \leqslant R$. Thus, in place of the radiation condition [which we imposed as $\rho \to +\infty$ for the convex surface, and led to the choice of $H^{(2)}_{\nu kR}(k\rho_>)$ in (10.11)], we must impose a condition of finiteness of the total field at $\rho = 0$ (there being no actual physical scatterer there).

The Fourier representation[6] for the Green's function G_3 of the concave surface is now obtained as in section 10.2:

$$G_3 = -\frac{jkR}{8}\int_{C_3} e^{-jk\nu|s-s'|}J_{\sqrt{\nu^2}kR}(k\rho_<)\{H^{(2)}_{\sqrt{\nu^2}kR}(k\rho_>) - Q(\sqrt{\nu^2},R)J_{\sqrt{\nu^2}kR}(k\rho_>)\}d\nu \tag{10.13}$$

$$s = R\phi;\ \rho_{\gtrless} = \begin{matrix}\max\\ \min\end{matrix}\{\rho,\rho'\};\ \mathrm{Re}\sqrt{\nu^2} \geqslant 0$$

$$Q(\nu,R) = \frac{H^{(2)'}_{\nu kR}(kR) - \gamma_0 H^{(2)}_{\nu kR}(kR)}{J'_{\nu kR}(kR) - \gamma_0 J_{\nu kR}(kR)} \tag{10.14}$$

The integration contour C_3 is shown in Fig. 10.4.

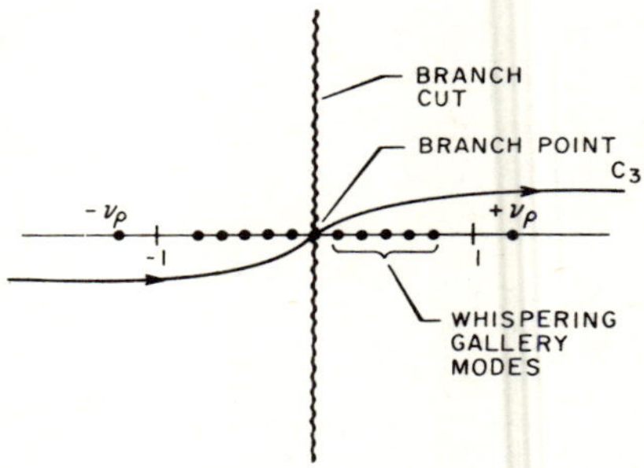

Fig. 10.4 ν-plane contour for concave surface

Two main features distinguish (10.13) from (10.11). In the first place, the presence of $\sqrt{\nu^2}$ introduces a pair of branch cuts* from the origin of the complex ν-plane, as shown in Fig. 10.4. These arise from the requirement that G_3 be finite at $\rho = 0$, since Bessel functions whose index has a negative real part become infinite there. Secondly, the remaining singularities in the ν-plane are the poles of Q, which differ from those of P. As for the convex surface, one of the pairs of roots, $\pm\nu_p$, corresponds to the surface modes $\pm\nu_0$ on the straight surface. As can be inferred from known properties of Bessel functions,[7] all the other poles lie on the real axis between ± 1. The number of pairs of these poles, which are known as whispering gallery poles, is large (roughly kR/π) but finite.

If C_3 is deformed into the lower half-plane, the pole residues at ν_p and $\nu_1, \ldots, \nu_N$ (the whispering gallery modes) give the discrete modal component, and the branch cut integral around the line from

* $\sqrt{\nu^2}$ may be thought of as $\lim_{\epsilon \to 0} (\nu^2 + \epsilon^2)^{1/2}$, with branch points at $\nu = \pm j\epsilon$

0 to $-j\infty$ gives the continuous spectrum component. This mode structure is again different (though not so much so as in the convex case) from that of the straight surface, and matching with the modes of a straight section is not easily accomplished using perturbation procedures. As opposed to the creeping modes which quickly decay away from a source, however, the whispering gallery modes do not attenuate at all, and the amplitudes of individual modes are of interest in specific problems. For example, some structures proposed for use in integrated optics use whispering gallery modes as the only modes of operation.[8] An approximate field representation which simply mimics the form of the modal expansion on a straight section, such as will be derived for the convex surface in the next two sections, is therefore inadequate for many purposes. This question will be left open, and the remainder of this section devoted to describing the features of the whispering gallery mode representation.

The field of a whispering gallery mode, outside of a constant factor, can be seen from (10.13) to be $J_{\nu_n kR}(k\rho)$. The properties of this field were studied as early as 1910 by Lord Rayleigh.[9] The function $J_{\nu_n kR}(k\rho)$, as we have seen in earlier chapters for $H^{(2)}_{\nu kR}$, possesses a caustic at $\nu_n kR = k\rho_{t_n}$, which appears at $\rho_{t_n} < R$ since $\nu_n < 1$. Without going into the detailed derivation of the asymptotic behavior of the whispering gallery mode functions (which is similar to that carried out for the radiating surface wave in section 7.7 and Appendix D) the following general statements can be made:

(i) For $\rho > \rho_{t_n}$, the mode function is an oscillating standing wave, as follows from the Debye expansion of $J_{\nu kR}$;

(ii) Similarly, for $\rho < \rho_{t_n}$, the mode function is a *purely* exponentially decaying field away from the caustic.

A consequence of statement (ii) is that in contrast with the caustic of a radiating surface wave, the caustic of a whispering gallery mode is perfectly reflecting, and only an evanescent field penetrates beyond it (much like total internal reflection at a dielectric interface). The physical interpretation of the whispering gallery phenomenon is the eventual re-reflection of a ray by the concave surface, a feature which is, of course, absent on the convex surface. The lossless reactive surface and the inner caustic thus together form a waveguide which guides its whispering modes without attenuation along the vicinity of the surface, in addition to what has been called the surface mode at ν_p (which has no such caustic since $\nu_p > 1$). In more realistic structures, the outer surface which has here been modeled by a reactance, will not be perfectly reflecting, and so the whispering gallery modes will have some attenuation due to the mechanism described in Chapter 7.

10.4 Approximate surface mode for a convex reactive surface

It is necessary to make two distinct assumptions in the derivation, which will be made more quantitative later on. First, it is required that the radius of the curved section be large compared to a wavelength ($kR \gg 1$), and second, both source and observation points should be close to the surface ($x/R, x'/R \ll 1$). The latter is not very restrictive for our purposes, since the major portion of the field will be the surface wave, which is indeed closely confined near the surface. If, for example, we wished to investigate a line source located high above the surface (so that the surface wave is virtually not excited) these assumptions would require some modification.

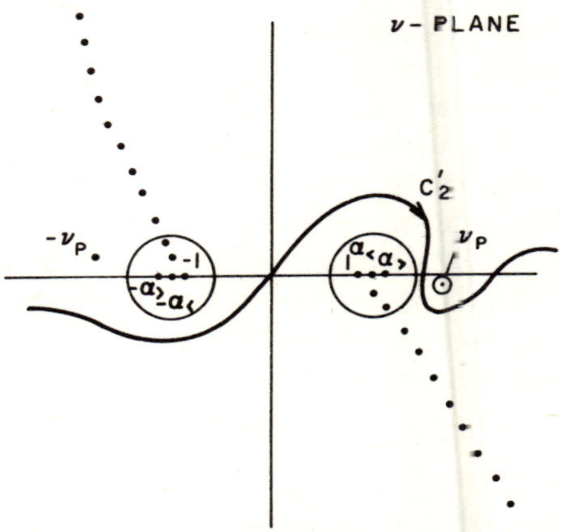

Fig. 10.5 Contour deformation, picking up surface-wave pole and avoiding $\nu = \pm 1, \pm \alpha_{\gtrless}$

If the surface-wave pole at ν_p is, by virtue of the above assumptions (and we will come back to this point), sufficiently to the right of the points $\nu = 1$, $\alpha_>$ and $\alpha_<$, where $\alpha_{\gtrless} = \rho_{\gtrless}/R$, we can deform C_2 into C_2' as shown in Fig. 10.5, in such a way that the residue at ν_p is picked up, while avoiding a region of radius on the order of $(kR)^{-2/3}$ about 1, $\alpha_>$ and $\alpha_<$. We have now split G_2 into a surface-wave part and a radiation part, but only the surface-wave part is in the form of a product of mode functions;

$$G_2 = G_{2s} + G_{2c}$$

where

$$G_{2s} = -j\frac{\gamma_0}{\nu_0}e^{-jk\nu_p|s-s'|}\,\Psi_s(x,R)\,\Psi_s(x',R) \qquad (10.15)$$

and

$$\Psi_s = \left\{ + \frac{j\nu_0 \pi kR}{4\gamma_0} \frac{H_{\nu_p}^{(1)'}(kR) + \gamma_0 H_{\nu_p kR}^{(1)}(kR)}{\frac{\partial}{\partial \nu}\left[H_{\nu kR}^{(2)'}(kR) + \gamma_0 H_{\nu kR}^{(2)}(kR)\right]_{\nu=\nu_p}} \right\}^{1/2} H_{\nu_p kR}^{(2)}(k\rho) \tag{10.16}$$

$$= \left\{ - \frac{\nu_0}{\gamma_0} \frac{H_{\nu_p kR}^{(2)}(kR)}{\frac{\partial}{\partial \nu}\left[H_{\nu kR}^{(2)'}(kR) + \gamma_0 H_{\nu kR}^{(2)}(kR)\right]_{\nu=\nu_p}} \right\}^{1/2} \frac{H_{\nu_p kR}^{(2)}(k\rho)}{H_{\nu_p kR}^{(2)}(kR)}$$

In obtaining (10.16), the Wronskian of $H^{(1)}$ and $H^{(2)}$ was used. The remaining term G_{2c} is the contribution from C_2' which will be dealt with shortly. $H_{\nu_p kR}^{(2)}(k\rho)$ is, of course, precisely the surface wave field used in sections 7.3 and 7.7 to investigate the local radiation condition on a two-dimensional structure. For use in waveguide junction problems, however, we are not particularly interested in the continuous radiation losses which must accumulate over a length of the curved section, and only affect the surface-wave field in an exponentially small fashion in kR near the surface.

Thus, the term outside the brackets in (10.16) is expanded as in Appendix C, but the exponentially small, locally growing part, which gives rise to the continuous radiation loss, will be dropped. The term inside the brackets is represented in the form

$$\frac{N(\nu_p)}{\frac{\partial}{\partial \nu}\left[D(\nu)\right]_{\nu=\nu_p}} \tag{10.17}$$

where, from the Debye form for $H^{(2)}$ (again neglecting the exponentially small part):

$$N(\nu) \sim 1 + \frac{1}{kR}\left[\frac{j}{24w^3}(3 + 2\nu^2)\right] + \ldots \tag{10.18}$$

$$D(\nu) \sim (\gamma_0 - jw) + \frac{1}{kR}\left[-\frac{1}{2w^2} + \frac{j}{24w^3}(\gamma_0 - jw)(3 + 2\nu^2)\right] \tag{10.19}$$

Now (10.17) is evaluated by the method of Appendix I; from (I.4) and (I.5) we get

$$\nu_p \sim \nu_0 + \nu_1/kR + \ldots \tag{10.20}$$

where

$$\nu_0 = (1 + \gamma_0^2)^{1/2}; \qquad \nu_1 = 1/(2\nu_0\gamma_0) \tag{10.21}$$

and by (I.6) the term in brackets in (10.16) is asymptotic to

$$\left\{1-\frac{1}{kR}\left[\frac{1+2\gamma_0^2}{2\nu_0^2\gamma_0^3}\right]+\ldots\right\}^{1/2}$$
$$\sim 1-\frac{1}{kR}\left[\frac{1+2\gamma_0^2}{4\nu_0^2\gamma_0^3}\right]+\ldots \tag{10.22}$$

On the other hand, from (C.13) and (10.20)

$$\frac{H^{(2)}_{\nu_p kR}(k\rho)}{H^{(2)}_{\nu_p kR}(kR)}\sim \exp\,[-kx(\nu_p^2-1)^{1/2}]\left\{1+\frac{kx}{2kR}\left[\frac{1}{\nu_p^2-1}+\frac{\nu_p^2 kx}{(\nu_p^2-1)^{1/2}}\right]\right\}$$
$$\sim \exp\,[-\gamma_0 kx]\left\{1+\frac{(kx)^2\nu_0^2}{2kR\gamma_0}\right\} \tag{10.23}$$

so that Ψ_s has the asymptotic expansion

$$\Psi_s\sim\Psi_{s0}(x)+\frac{1}{kR}\Psi_{s1}(x)+\ldots \tag{10.24}$$

where Ψ_{s0} is given in (10.6) and

$$\Psi_{s1}(x)\;=\;\exp\,(-\gamma_0 kx)\left[\frac{\nu_0^2(kx)^2}{2\gamma_0}-\frac{1+2\gamma_0^2}{4\nu_0^2\gamma_0^3}\right] \tag{10.25}$$

valid for $(kR\gamma_0^3)\gtrsim 1$, and kx finite.

Although higher order terms in (10.20) and (10.24) could, by the above method, be found in principle, it is important to note that they may also be found by means of the constructive method of Chapters 4–6 at least up to the arbitrary normalization discussed in Chapter 4. This approach is considerably simpler, but does not immediately give an estimate of the validity of the expansion. The normalization corresponding to Ψ_s as defined in (10.16) can be shown to be

$$\int_0^\infty [\Psi_s(x,R]^2\,\frac{R}{R+x}\,dx\;=\;\frac{\nu_0/\nu_p}{2k\gamma_0}, \tag{10.26}$$

This follows from the properties of the Hankel functions. The regions of validity of the corresponding expansions for closed waveguide modes in Chapters 4–6 may be derived in similar fashion.

10.5 Radiation quasimodes for a convex reactive surface

We have yet to obtain a representation for G_{2c} which resembles the second term of (10.5). Let us rewrite G_{2c} in the form

$$G_{2e} = -\frac{jkR}{4}\int_{C_2'} e^{-jk\nu|s-s'|} H_{\nu kR}^{(2)}(kR) J_{\nu kR}(kR) \left[\frac{H_{\nu kR}^{(2)}(k\rho_>)}{H_{\nu kR}^{(2)}(kR)}\right]$$
$$\cdot \left\{\frac{J_{\nu kR}(k\rho_<)}{J_{\nu kR}(kR)} - \frac{H_{\nu kR}^{(2)}(k\rho_<)}{H_{\nu kR}^{(2)}(kR)} \cdot \frac{H_{\nu kR}^{(2)}(kR)}{2J_{\nu kR}(kR)} [1 + P(\nu, R)]\right\} d\nu \qquad (10.27)$$

A single, convenient, continuous expansion for the integrand whose exponentially dominant and subdominant parts are correct over all of C_2' cannot be found; however, in the spirit of the foregoing approximation for the surface mode, we can use approximations whose *dominant* part is correct over all of C_2'. The Debye expansion for J and the Debye expansion for $H^{(2)}$ with subdominant part dropped fit this condition,[10, 11] if we specify the branch of $(1-y^2)^{1/2}$ (see Appendix C)

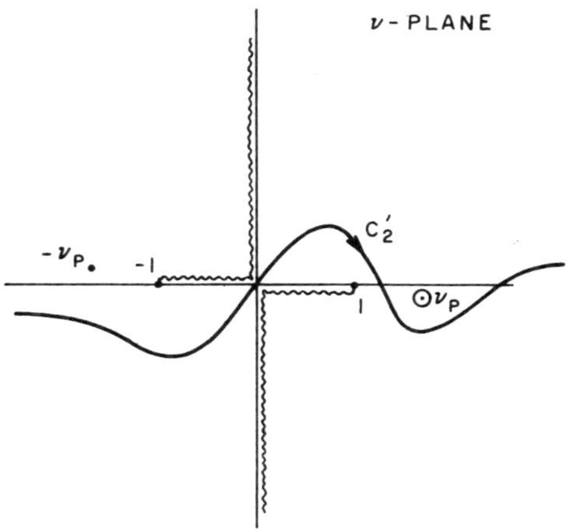

Fig. 10.6 ν-plane for G_2 after Debye and Taylor expansions. The Riemann sheet is Im $w < 0$

so that $\arg(1-y^2)^{1/2} = 0$ for $|y| < 1$ and $\arg(1-y^2)^{1/2} = +\pi/2$ for $|y| > 1$. Put another way, we must draw branch cuts in the ν-plane defining w so that the cut from $\nu = +1$ lies below C_2' and the cut from $\nu = -1$ lies above C_2'. One such set of cuts (and the obvious choice, considering Fig. 10.2) is shown in Fig. 10.6. It should be noted that the discarding of the subdominant part of $H^{(2)}$ has caused the creeping wave poles to disappear and be replaced by the branch cuts. Only the one pair of branch cuts appears if the Debye-Taylor expansions from Appendix C are used:

$$\frac{H_{\nu kR}^{(2)}(k\rho)}{H_{\nu kR}^{(2)}(kR)} \sim \exp\,[-jkwx] \left\{1 + \frac{kx}{2kR}\left[-\frac{1}{w^2} - \frac{j\nu^2 kx}{w}\right]\right\} \qquad (10.28)$$

$$\frac{J_{\nu kR}(k\rho)}{J_{\nu kR}(kR)} \sim \exp\,[+jkwx]\left\{1+\frac{kx}{2kR}\left[-\frac{1}{w^2}+\frac{j\nu^2 kx}{w}\right]\right\} \quad (10.29)$$

By the choice of branch cuts, w is now defined just as in (10.3).

The remainder of the integrand of (10.27) is approximated using the Debye expansions,

$$\frac{1}{\pi kRw}\,[S(\nu,R)]^2 \equiv H^{(2)}_{\nu kR}(kR)\,J_{\nu kR}(kR) \sim \frac{1}{\pi kRw}\,\{1+0(kR)^{-2}\} \quad (10.30)$$

$$\Gamma(\nu,R) \equiv -\frac{H^{(2)}_{\nu kR}(kR)}{2J_{\nu kR}(kR)}\,[1+P(\nu,R)] = \left[-j\frac{J'_{\nu kR}(kR)}{J_{\nu kR}(kR)}-j\gamma_0\right]$$

$$\cdot\left[j\frac{H^{(2)\prime}_{\nu kR}(kR)}{H^{(2)}_{\nu kR}(kR)}+j\gamma_0\right]^{-1} \equiv [L(\nu,R)]\cdot[M(\nu,R)]^{-1} \quad (10.31)$$

The asymptotic expansions of L and M are obtained with the help of (C.15) and the dominant part of (C.17) in Appendix C:

$$L(\nu,R) \sim w - j\gamma_0 + \frac{j}{2kRw^2} \quad (10.32)$$

$$M(\nu,R) \sim w + j\gamma_0 - \frac{j}{2kRw^2} \quad (10.33)$$

We are now in a position to deform C_2' over the lower branch cut in Fig. 10.6. It can be seen from the above expansions that besides the branch cut, the only other possible singularity is a high-order negative power of w contributing at $\nu = 1$. In fact, this problem does not arise with the terms we have retained, but higher-order terms may require interpretation of the branch cut integral through $\nu = 1$ (i.e, through $w = 0$) in the sense of generalized functions: a Cauchy principal value or Hadamard finite part.[12] The integral is known to be convergent before the contour deformation; in this way, we are assured that the branch cut integrals will be finite as well.

Insofar as the asymptotic relations (10.28) – (10.31) are concerned, making the substitution $w \to -w$ effects the transformations

$$H^{(2)}_{\nu kR}(k\rho)/H^{(2)}_{\nu kR}(kR) \longleftrightarrow J_{\nu kR}(k\rho)/J_{\nu kR}(kR)$$

$$S(\nu,R) \longrightarrow S(\nu,R)$$

$$\Gamma(\nu,R) \longrightarrow 1/\Gamma(\nu,R)$$

All of the expressions above are to be understood as formal shorthand for the asymptotic relations (10.28) – (10.31). The branch cut integral can now be transformed into one over positive w only

$$G_{2c} \sim -\frac{j}{4\pi}\int_0^\infty \frac{[S(\nu,R)]^2}{\Gamma(\nu,R)}\left\{\frac{J_{\nu kR}(k\rho)}{J_{\nu kR}(kR)} + \Gamma(\nu,R)\frac{H^{(2)}_{\nu kR}(k\rho)}{H^{(2)}_{\nu kR}(kR)}\right\}\left\{\frac{J_{\nu kR}(k\rho')}{J_{\nu kR}(kR)} + \Gamma(\nu,R)\frac{H^{(2)}_{\nu kR}(k\rho')}{H^{(2)}_{\nu kR}(kR)}\right\}\cdot e^{-jk\nu|s-s'|}\frac{dw}{\nu} \tag{10.34}$$

where it is still understood that $S(\nu,R)$, $\Gamma(\nu,R)$ etc. stand for their asymptotic expansions. Be regrouping the integrand, we recover a representation analogous to the integral in (10.5):

$$G_{2c} \sim -\frac{j}{2}\int_0^\infty \Psi_c(w,x,R)\,\Psi_c(w,x',R)\,e^{-jk\nu|s-s'|}\frac{dw}{\nu N^2(w)} \tag{10.35}$$

Here ν and $N^2(w)$ are given in (10.5), while

$$\Psi_c(w,x,R) \sim \frac{S(\nu,R)}{2}\left\{M(\nu,R)\frac{J_{\nu kR}(k\rho)}{J_{\nu kR}(kR)} + L(\nu,R)\frac{H^{(2)}_{\nu kR}(k\rho)}{H^{(2)}_{\nu kR}(kR)}\right\}\times \left\{\frac{w+j\gamma_0}{M(\nu,R)}\cdot\frac{w-j\gamma_0}{L(\nu,R)}\right\}^{1/2} \tag{10.36}$$

which according to (10.30) – (10.33) is given as

$$\sim \Psi_{c0}(w,x) + \frac{1}{kR}\Psi_{c1}(w,x) + \ldots \tag{10.37}$$

Once again, the leading term Ψ_{c0} turns out to be the straight guide limit (10.7), and higher terms can be gotten by inserting all the previous expansions into (10.36). For example,

$$\Psi_{c1}(w,x) = \frac{1}{2}\left\{(w+j\gamma_0)\,e^{jkwx}\left[-\frac{j}{2w(w^2+\gamma_0^2)} + \frac{kx}{2}\left(-\frac{1}{w^2} + \frac{j\nu^2kx}{w}\right)\right] + (w-j\gamma_0)\,e^{-jkwx}\left[+\frac{j}{2w(w^2+\gamma_0^2)} + \frac{kx}{2}\left(-\frac{1}{w^2} - \frac{j\nu^2kx}{w}\right)\right]\right\} \tag{10.38}$$

To find higher order terms of (10.36) by this method would probably be even more tedious than for (10.24). Again it turns out, however, that the constructive methods of Chpaters 4–6 can be used to determine Ψ_c up to an arbitrary normalization. The normalization for Ψ_c is found to be

$$\int_0^\infty \Psi_c(w,x,R)\,\Psi_c(w',x,R)\frac{R}{R+x}\,dx \sim N^2(w)\delta(w-w')/k \tag{10.39}$$

This can be accomplished using (10.36) and the known properties of cylinder functions, and is an interesting, nontrivial exercise in the use of generalized functions. In addition, we have

$$\int_0^\infty \Psi_c(w,x,R)\,\Psi_s(x,R)\,\frac{R}{R+x}\,dx \sim 0 = 0((kR)^{-N}) \quad (10.40)$$

for any positive N. The 'orthogonality' statements (10.39) and (10.40) should be understood in the following sense: if (10.39) is multiplied by some suitable function $\phi(w')$ and integrated from $w' = 0$ to ∞, the result is asymptotic (as far as inverse powers of kR are concerned) to $N^2(w)\phi(w)/k$. Relation (10.40) may be thought of in the ordinary asymptotic sense.

10.6 Construction of quasimodes for a convex reactive surface

The quasimodes (10.24) and (10.37) can be constructed directly from the differential equation in the manner of Chapters 4–6, along with a few additional, physically plausible conditions. The wave equation for the curved section can be written as

$$(x+R)\frac{\partial}{\partial x}\left[(x+R)\frac{\partial \Psi}{\partial x}\right] + R^2\frac{\partial^2 \Psi}{\partial s^2} + k^2(x+R)^2\Psi = 0 \quad (10.41)$$

The surface wave will be considered first. Assuming the expansions (10.20) and (10.24) for ν_p and Ψ_s, with ν_i and Ψ_{si} to be determined, they are substituted into (10.41), and like powers of $(kR)^{-1}$ are equated to zero to give the recurrent sequence of equations

$$\Psi''_{s0} + k^2(1-\nu_0^2)\,\Psi_{s0} = 0 \quad (10.42)$$

$$\Psi''_{s1} + k^2(1-\nu_0^2)\,\Psi_{s1} = -2kx\Psi''_{s0} - k\Psi'_{s0} + 2k^2(\nu_0\nu_1 - kx)\,\Psi_{s0} \quad (10.43)$$

and so on, subject to the boundary conditions (see (10.1))

$$[\Psi'_{si} + k\gamma_0\Psi_{si}]_{x=0} = 0; \qquad i = 0, 1, 2, \ldots \quad (10.44)$$

If as $x \to \infty$ it is required that exponentially growing terms be absent from the Ψ_{sj}, equations (10.42) and (10.44) are easily solved for Ψ_{s0} and ν_0 as given in (10.6) and (10.21). Equations (10.43)–(10.44) are likewise solved for ν_1 and Ψ_{s1} (the latter up to an arbitrary multiple of Ψ_{s0}) just as in Chapter 4. As previously noted, this arbitrariness can be eliminated by enforcing a normalization [in this case, relation (10.26)], and (10.21) and (10.25) are the end result. Higher orders are constructed similarly, and clearly the process is much less tedious than that of section 10.4.

In like manner, the continuous quasimodes are obtained by assuming the representation (10.37), but w is still taken to be real and positive, assuring no exponentially growing solutions as $x \to \infty$. Inserting (10.37) into (10.41) leads to a set of recursion equations similar to (10.42) – (10.43), which can be solved in the same way. Arbitrary multiples of Ψ_{c0} which arise at each step of the construction are determined by enforcing the normalization (10.39). In this situation, the integrals obtained must be treated in the sense of generalized functions. The results for Ψ_{c0} and Ψ_{c1} are precisely as given by (10.7) and (10.38).

10.7 Discussion of quasimode representation

The mode functions Ψ_s and Ψ_c, since they provide an asymptotic basis for representing the field due to a line source over the convext surface, are asymptotically 'complete' in that an arbitrary field concentrated in or near the boundary layer may be represented by them. The limitations imposed by the derivation are that [see (10.25)]:

$$\gamma_0^3 kR \gtrsim 1$$

and

$$kx \lesssim (\gamma_0 kR)^{1/2}$$

The boundary layer thus specified does not allow $x \to \infty$, but does permit kx to be large enough to consider a 'far field' within the boundary layer. We should emphasize once again that the Ψ_c are *not* asymptotic expansions of proper modes of the curved section (as is Ψ_s), and strictly speaking, we denote this by referring to them as 'quasimodes', which taken together can provide an asymptotically correct basis on which to expand a field. With this understanding, we will often drop the distinction when solving the curvature discontinuity problem in the next chapter.

References

1 Shevchenko, V.V.: 'Continuous transitions in open waveguides' (Golem Press, Boulder, 1971), pp. 22-77

2 Felsen, L.B., and Marcuvitz, N.: 'Radiation and scattering of waves' (Prentice Hall, Englewood Cliffs, 1973), pp. 311–312

3 *Ibid.*, pp. 685–691

4 Keller, J.B., Rubinow, S.I., and Goldstein, M.: 'Zeros of Hankel functions and poles of scattering amplitudes', *J. Math. Phys.*, Vol. 4, No. 6, June 1963

5 Buchal, R.N., and Keller, J.B.: 'Boundary layer problems in diffraction theory'. *Comm. Pure Appl. Math.*, Vol. 13, pp. 85–114, 1960

6 Wasylkiwsky, W.: 'Diffraction by a concave perfectly conducting cylinder', *IEEE Trans. Ant. Prop.*, Vol. 23, pp. 480–492, 1975

7 Coulomb, J.: 'Sur les zéros des fonctions de Bessel considérés comme fonction de l'ordre', *Bull. Sci. Math.*, (1 re partie) sér. 2, Vol. 60, pp. 297–302, 1930

8 Sheem, S., and Whinnery, J.R.: 'Modes of a curved surface waveguide for integrated optics', *Wave Electron.*, Vol. 1, pp. 105–116, 1975

9 Lord Rayleigh (J.W. Strutt): 'The problem of the whispering gallery', *Phil. Mag.*, Vol. 20, pp. 1001–1004, 1910

10 Ref. 2, pp. 710–712

11 Nussenzveig, H.M.: 'High-frequency scattering by an impenetrable sphere', *Annals of Physics*, Vol. 34, pp. 23–95, 1965

12 Jones, D.S.: 'Generalised functions' (McGraw-Hill, London, 1966), pp. 89–92

Chapter 11

Scattering of a surface wave at a curvature discontinuity

11.1 Mode-match solution for the junction of straight and curved impedance surfaces

An application of the quasimode representation found in the previous chapter is to the scattering of a surface wave at the junction between a straight and a convex reactive surface. The configuration is shown in Fig. 11.1. Let us assume that the surface wave supported by the straight section is incident upon the junction from the left. The procedure is formally the same as that of Chapter 5.

On the left of the junction ($y < 0$), the field is represented as

$$E_z = [e^{-jk\nu_0 y} + \Omega e^{+jk\nu_0 y}]\ \Psi_{s0}(x) + \int_0^\infty B_-(w)\,\Psi_{c0}(w,x)\,e^{jk\nu y}\,dw \tag{11.1}$$

where Ψ_{s0} and Ψ_{c0} are given by (10.6) and (10.7) and

$$\begin{aligned} \nu &= (1-w^2)^{1/2} & 0 \leqslant w \leqslant 1 \\ &= -j(w^2-1)^{1/2} & 1 \leqslant w < \infty \end{aligned}$$

Similarly the field on the curved section ($s > 0$) is put in the form

$$E_z = T\Psi_s(x,R)\,e^{-jk\nu_p s} + \int_0^\infty B_+(w)\Psi_c(w,x,R)\,e^{-jk\nu s}\,dw \tag{11.2}$$

where Ψ_s and Ψ_c are given by (10.20) and (10.32). In these representations Ω is the reflection coefficient of the surface wave, T its transmission coefficient, while $B_\pm(w)$ will later be related to the pattern of the scattered radiation field.

From Maxwell's equations, we have

$$\begin{aligned} H_x &\propto \partial E_z/\partial y & (y<0) \\ H_\rho &\propto [R/(R+x)]\,\partial E_z/\partial s & (s>0) \end{aligned}$$

and so enforcing continuity of tangential E and H at the junction ($y = s = 0$) results in

$$(1+\Omega)\Psi_{s0}(x)+\int_0^\infty B_-(w)\Psi_{c0}(w,x)\,dw = T\Psi_s(x,R) + \int_0^\infty B_+(w)\Psi_c(w,x,R)\,dw \tag{11.3}$$

$$\nu_0(1-\Omega)\Psi_{s0}(x)-\int_0^\infty \nu B_-(w)\Psi_{c0}(w,x)\,dw = \frac{R}{R+x}\left\{\nu_p T\Psi_s(x,R) + \int_0^\infty \nu B_+(w)\Psi_c(w,x,R)\,dw\right\} \tag{11.4}$$

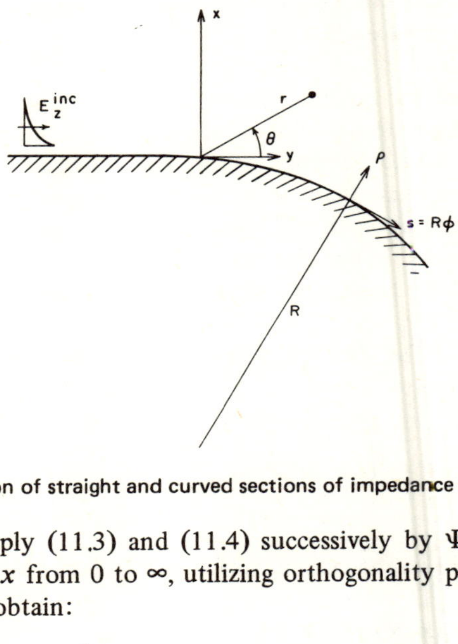

Fig. 11.1 Junction of straight and curved sections of impedance surface

We now multiply (11.3) and (11.4) successively by Ψ_{s0} and Ψ_{c0} and integrate over x from 0 to ∞, utilizing orthogonality properties (10.8) –(10.10) and obtain:

$$(1+\Omega)\frac{1}{2k\gamma_0} = TI_1(R)+\int_0^\infty B_+(w)I_2(w,R)dw \tag{11.5}$$

$$\nu_0(1-\Omega)\frac{1}{2k\gamma_0} = \nu_p TI_3(R)+\int_0^\infty \nu B_+(w)I_4(w,R)dw \tag{11.6}$$

$$B_-(w)N^2(w)/k = TI_5(w,R)+\int_0^\infty B_+(w')I_6(w,w',R)dw' \tag{11.7}$$

$$-\nu B_-(w)N^2(w)/k = \nu_p TI_7(w,R)+\int_0^\infty \nu' B_+(w')I_8(w,w',R)dw' \tag{11.8}$$

I_1–I_8 are shorthand for various mode-product integrals which can be evaluated in inverse power series in R directly from the definitions of Ψ_s, Ψ_{s0}, Ψ_c and Ψ_{c0}:

$$I_1(R) = \int_0^\infty \Psi_{s0}(x)\Psi_s(x,R)dx \sim \frac{1}{2k\gamma_0} + \frac{1}{kR}\left[\frac{1}{8k\nu_0^2}\right] + \ldots \quad (11.9)$$

$$I_2(w,R) = \int_0^\infty \Psi_{s0}(x)\Psi_c(w,x,R)dx \sim \frac{1}{kR}\left[\frac{w(\nu^2+\nu_0^2)}{k(w^2+\gamma_0^2)^2}\right] + \ldots \quad (11.10)$$

$$I_3(R) = \int_0^\infty \left(\frac{R}{R+x}\right) \Psi_{s0}(x)\Psi_s(x,R)dx \sim \frac{1}{2k\gamma_0} - \frac{1}{kR}\left[\frac{2+\gamma_0^2}{8k\nu_0^2\gamma_0^2}\right] + \ldots \quad (11.11)$$

$$I_4(w,R) = \int_0^\infty \left(\frac{R}{R+x}\right)\Psi_{s0}(x)\Psi_c(w,x,R)dx \sim \frac{1}{kR}\left[\frac{2w\nu_0^2}{k(w^2+\gamma_0^2)^2}\right] + \ldots \quad (11.12)$$

$$I_5(w,R) = \int_0^\infty \Psi_s(x,R)\Psi_{c0}(w,x)dx \sim \frac{1}{kR}\left[-\frac{2w\nu_0^2}{k(w^2+\gamma_0^2)^2}\right] + \ldots \quad (11.13)$$

$$I_6(w,w',R) = \int_0^\infty \Psi_{c0}(w,x)\Psi_c(w',x,R)dx$$
$$\sim N^2(w)\delta(w-w')/k + 0(1/kR) \quad (11.14)$$

$$I_7(w,R) = \int_0^\infty \left(\frac{R}{R+x}\right)\Psi_s(x,R)\Psi_{c0}(w,x)dx \sim \frac{1}{kR}\left[-\frac{w(\nu^2+\nu_0^2)}{k(w^2+\gamma_0^2)^2}\right] + \ldots \quad (11.15)$$

$$I_8(w,w',R) = \int_0^\infty \left(\frac{R}{R+x}\right) \Psi_{c0}(w,x)\Psi_c(w',x,R)dx$$
$$\sim N^2(w)\delta(w-w')/k + 0(1/kR) \quad (11.16)$$

The functions I_6 and I_8 are, of course, to be considered as generalized functions.

A perturbation solution of (11.5)–(11.8) for large kR can be found by assuming asymptotic expansions of the form

$$T \sim T^{(0)} + \frac{1}{kR}T^{(1)} + \frac{1}{(kR)^2}T^{(2)} + \ldots \quad (11.17)$$

for the four unknowns $(T, R, B_\pm(w))$ and inserting them, along with (11.9)–(11.6) and the expansion (10.6) for ν_p into (11.5)–(11.8). Like inverse orders of kR are then matched, and it is straightforward to obtain the leading terms:

$$T = 1 + 0[1/(kR)^2] \quad (11.18)$$

$$\Omega = \frac{1}{kR}\left[\frac{\gamma_0}{4\nu_0^2}\right] + 0[1/(kR)^2] \tag{11.19}$$

$$B_\pm(w) = \frac{1}{kR}\left[\frac{w\nu_0(\nu_0 \pm \nu)^2}{2\nu N^2(w)(w^2+\gamma_0^2)^2}\right] + 0[1(kR)^2] \tag{11.20}$$

The solution, to this order, is in agreement with Shevchenko[1], who attacks the problem in terms of continuous coupling of straight guide modes on the curved section. The present approach permits separate consideration of a single junction, since it is unnecessary to terminate the curved section with another straight section, and moreoever, the discussion of the previous chapter on the validity of the representation can be extended to the solution of this problem as well, emphasizing its limitation to the boundary layer.

11.2 Field near the junction

For $y < 0$, the field at the surface of the straight section ($x = 0$) is, from (11.1), (11.19) and (11.20):

$$E_z \sim e^{-jk\nu_0 y} + \Omega e^{+jk\nu_0 y} + \frac{\nu_0}{\pi kR}\int_0^\infty \frac{w^2 e^{+jk\nu y}}{\nu(\nu_0+\nu)^2(w^2+\gamma_0^2)}\,dw \tag{11.21}$$

As $y \to 0^-$ the surface wave component remains well-behaved, while the radiation component is dominated by the value of the integrand for large w. Rewritten as a ν integral, the latter is

$$-\frac{\nu_0}{\pi kR}\left\{\int_1^0 + \int_0^{-j\infty}\right\}\frac{w e^{+jk\nu y}}{(\nu_0+\nu)^2(\nu_0^2-\nu^2)}\,d\nu \tag{11.22}$$

The non-exponential part of the integrand behaves as ν^{-3} as $|\nu| \to \infty$, and so, in addition to some analytic terms, there will be a field contribution which behaves like the exponential integral of third order:[2]

$$E_3(-ky) = \int_1^\infty \frac{e^{+kyt}}{t^3}\,dt \tag{11.23}$$

As was found in the closed waveguide problem considered in Chapter 5, the field remains finite at $y = 0$, but its second derivative is logarithmically singular: $d^2E_3(x)/dx^2 = E_1(x) \sim -\log x$. Such strikingly similar behavior is suggestive of the existence of a generalized 'edge criterion' similar to that which governs the singularity of fields at sharp edges, which would apply to any nonanalytic point of a boundary and specify the degree of nonanalyticity of the field there. There is a hint of such a thing in the conformal mapping investigation of a somewhat different problem by Hong and Goodrich,[3] but this point will not be pursued further here.

11.3 Pattern of the scattered field

The far-field pattern of the scattered field is found by inserting $B_{\pm}(w)$ from (11.20) into (11.1) and (11.2), and evaluating the integrals by the saddle-point technique. In general, assuming $B_{\pm}(w)$ to be a suitable analytic function, the field on a straight section, given by the expression

$$E_z = \int_0^{\infty} B_{\pm}(w)\,\psi_{co}(w, x)\, e^{\mp jk\nu y}\, dw$$

will accordingly take the far-field form

$$E_z \sim \left(\frac{\pi}{2kr}\right)^{1/2} F(\theta)\, e^{-jkr+j\pi/4} \qquad (11.24)$$

where

$$F(\theta) = \pm \cos\theta\,(\sin^2\theta + \gamma_0^2)^{1/2} B_{\pm}(w)\Big|_{\substack{w=\sin\theta \\ \nu=\pm\cos\theta}} e^{-iq(\theta)} \qquad (11.25)$$

and $\tan\theta = x/y$, $\tan q(\theta) = \gamma_0/\sin\theta$ and $kr = k(x^2 + y^2)^{1/2}$ is large compared to unity.

Similarly, on the curved section,

$$E_z = \int_0^{\infty} B_{\pm}(w)\,\psi_c(w, x, R)\, e^{\mp jk\nu s}\, dw$$

$$\sim \left(\frac{\pi}{2k\tilde{r}}\right)^{1/2} F(\tilde{\theta})\, e^{-jk\tilde{r}+j\pi/4} \qquad (11.26)$$

where F is given by (11.25), but $\tan\tilde{\theta} = x/s$, and $k\tilde{r} = k(x^2 + s^2)^{1/2}$ is large compared to unity (but kx must still be small enough compared to kR to remain within the boundary layer). Unless $\tilde{\theta}$ is close to 0 or π, $k\tilde{r} \ll kR$ in order to stay in the boundary layer, and so for the curvature discontinuity,

$$\begin{aligned} k\tilde{r} &= kr + 0(1/kR) \\ \tilde{\theta} &= \theta + 0(1/kR) \end{aligned} \qquad (11.27)$$

Near $\tilde{\theta} = 0$ (i.e., the surface of the curved section) the approximations (11.27) break down [and for that matter the saddle point evaluation of (11.26) must be carried to higher order, because in the present case $F(0) = 0$]. In general, a very complicated pattern is to be expected here, in view of the transition from an essentially junction-diffracted field to one diffracted by the curvature of the surface for larger ks.

Thus by substituting from (11.20), and taking (11.27) into account, we have for the radiation pattern of the curvature discontinuity

$$|F(\theta)| = \frac{\nu_0 \sin\theta}{\pi k R(\nu_0 - \cos\theta)^2(\nu_0^2 - \cos^2\theta)^{1/2}} \tag{11.28}$$

The single lobe of this pattern has a maximum at an angle θ_0 which is found by taking the derivative of (11.28) and setting it equal to zero. In the special case $\gamma_0/\nu_0 \ll 1$, which can be thought of as approximating, at least in the sense of weak guidance, a more physical optical waveguide, this maximum is located at $\theta_0 \simeq \gamma_0/2$, and thus is strongly endfire with radiation in the backward direction nearly absent. Clearly, when two or more discontinuities are present on the same structure, the fields scattered from each will interfere to produce a much more complicated overall radiation pattern.

We will close this section with two final remarks. First it is interesting to note than an analysis by Kaminetzky and Keller[4] of the diffraction of an incident *plane wave* at a curvature discontinuity gives an expression for the scattered field identical to the present one if, for the angle of incidence α, we formally insert the complex value with $\cos\alpha = \nu_0$ appropriate to an incident surface wave. Second, it should be possible to continue the radiation pattern outside of the boundary layer by appealing to arguments involving the induction theorem and equivalent magnetic and electric surface currents as in section 8.4. In contrast to the latter method, however, the field on the surface is *not* assumed to change only in the phase velocity of the surface wave, but having solved the junction problem it is known also that contributions arise from distortion of the surface wave on the curved section, conversion into radiation modes, and so on. Since these last were not considered in section 8.4, it is not surprising that different (although qualitatively similar) patterns should be obtained.

11.4 Power scattered at the junction

To the extent that the orthogonality statements (10.22), (10.34) and (10.35) are valid (that is, in an asymptotic sense), they imply, as do (10.8) – (10.10) for the straight section, that each surface or radiation mode carries power without conversion into other modes on a uniform section of guide. In fact, as a result of the normalization chosen for Ψ_s and Ψ_c, either of the surface wave fields

$$E_z = A_\pm \Psi_{s0}(x) e^{\mp jk\nu_0 y} \quad \text{or} \quad A_\pm \Psi_s(x) e^{\mp jk\nu_p s}$$

carries the time average power

$$P_s^{\pm} = \frac{1}{2}\,\mathrm{Re}\int_0^{\infty} E \times H^* \cdot a_{y,s}dx \sim \pm\frac{\nu_0/\gamma_0}{4\omega\mu_0}\,|A_{\pm}|^2 \qquad (11.29)$$

if the continuous attenuation of the surface mode on the curved section is neglected (a minus sign indicates power being carried in the $-y$ or $-s$ direction). Similarly, the collection of radiation modes

$$E_z = \int_0^{\infty} B_{\pm}(w)\,\psi_{c0}(w,x)e^{\mp jk\nu y}dw \text{ or } \int_0^{\infty} B_{\pm}(w)\,\psi_c(w,x,R)e^{\mp jk\nu s}dw$$

carries the time average power

$$P_c^{\pm} \sim \pm\frac{\pi}{4\omega\mu_0}\int_0^1 \nu(w^2+\gamma_0^2)\,|B_{\pm}(w)|^2\,dw \qquad (11.30)$$

Therefore, the power lost to the reflected surface wave at the junction [omitting the factor $(4\omega\mu_0)^{-1}$ everywhere] is

$$P_{refl} = \gamma_0/(16k^2R^2\nu_0^3) \qquad (11.31)$$

while that lost to radiation is

$$\begin{aligned} P_{rad} &= \frac{2\nu_0^2}{\pi k^2R^2}\int_0^1 \frac{w^2(\nu_0^4+6\nu_0^2\nu^2+\nu^4)}{\nu(w^2+\gamma_0^2)^5}\,dw \\ &= \frac{\nu_0}{16\gamma_0k^2R^2}\left[\frac{\nu_0^2(\nu_0^2+4)}{\gamma_0^6}-\frac{\gamma_0^2}{\nu_0^4}\right] \end{aligned} \qquad (11.32)$$

Thus the total power loss at the junction is, to this order in kR,

$$P_{scat} = P_{rad}+P_{refl} = \frac{\nu_0}{16\gamma_0k^2R^2}\,\frac{\nu_0^2(\nu_0^2+4)}{\gamma_0^6} \qquad (11.33)$$

Since the incident power in the surface wave is $P_{inc} = \nu_0/\gamma_0$, the relative power loss is

$$P_{scat}/P_{inc} = \delta^2(\nu_0^2+4)/16 \qquad (11.34)$$

where

$$\delta = \nu_0/(\gamma_0^3kR) \lesssim 1. \qquad (11.35)$$

It is of interest to compare this power loss with that due to continuous radiation of the surface wave around the bend. For a total angle Φ of bend, this loss is given by

$$P_{surf}/P_{inc} = 1-\exp(-2\alpha_s kR\Phi) \qquad (11.36)$$

where α_s is the normalized attenuation constant and is determinable by the methods of Chapters 7–9. We shall merely quote the result[5]

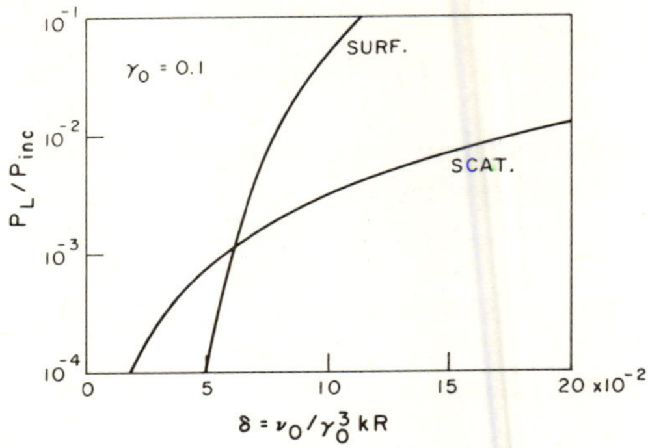

Fig. 11.2 Power lost by the surface-wave in a bend of $\Phi = 30°$, and the scattered power at the curvature discontinuity

$$\alpha_s kR\Phi = (\Phi/\gamma_0\delta) \exp [-1/\nu_0^2 - 2/(3\nu_0\delta)] \qquad (11.37)$$

if $\gamma_0/\nu_0 \ll 1$. In Fig. 11.2, (11.34) and (11.36) are compared for a bend angle of $\Phi = 30°$. Evidently, for extremely small ($\lesssim .05$) values of δ, the scattering losses will dominate, whereas as δ increases, the exponential dependence of (11.36) rapidly causes the attenuation of the surface wave to predominate. Because $\gamma_0 = 0.1$ used in Fig. 11.2 would be quite typical of an optical waveguide, either type of radiation could clearly be the more important and care must be exercised in designing curved open waveguide devices using either loss mechanism alone.

11.5 Concluding remarks

As mentioned in section 11.2, the behavior of the scattered field near the curved surface as *ks* increases is expected to involve a complex transition from the junction-diffracted field to a field dominated by a few creeping waves. In the present context, the total field near the surface is dominated by the surface wave component, and the presence or absence of creeping waves is irrelevant. It seems clear, however, that the radiation quasimodes cannot be expected to duplicate the behavior of the creeping waves, and in fact do not provide an asymptotic representation of the field in the deep shadow if the surface wave is absent.[6,7]

Although this is a minor point for the convex problem, the analogous situation for the concave problem presents more serious difficulties.

In the vicinity of the curvature discontinuity, before the scattered field has had a chance to 'see' the concavity of the curved section and suffer re-reflections, it seems intuitively clear that the field should be obtainable as for the convex case, and that the result will differ only by a change of sign in R. Once the re-reflections have become significant, a number of whispering gallery modes (see section 10.4) will be apparent.

There seem to be but few alternatives for dealing with this problem. An attempt could be made to use the whispering gallery modes themselves to expand the field at the junction (much the same as was done by Wait[8] for the convex surface). This formulation, however, does not easily lend itself to finding the backscattered field, and the excitation coefficients are integrals not expressible in elementary functions nor in ordinary special functions, so while it might be most suitable for a computer solution, little physical insight can be gained from it. Another approach could be taken based upon a study of the transition of the 'corrected flat-surface' field to the field diffracted by the curved surface. This transition has been studied by a number of methods on both concave and convex surfaces for the point source problem,[6,7,9–11] and could profitably be investigated for the present problem using matched asymptotic expansions.

It seems evident that the classes of problems found in this book do not begin to exhaust the possibilities of the methods outlined here and such extensions as proposed above. Many practical problems exist which combine features of waveguides with those of diffraction problems, and can be analyzed by these methods. Future developments will show which of these methods are the most suitable for the different classes of problems encountered in practical situations.

References

1 Shevchenko, V.V.: 'Continuous transitions in open waveguides' (Golem Press, Boulder, 1971) pp. 65–77 (a number of misprints occur in this section; in particular, the sign of the reflection coefficient is reversed)

2 Abramowitz, M., and Stegun, I.A. (ed.): 'Handbook of mathematical functions' (Applied Mathematics Series 55) Washington, DC, NBS, June 1964, p. 228

3 Hong, S., and Goodrich, R.F.: 'Application of conformal mapping to scattering and diffraction problems' in 'Electromagnetic wave theory' (J. Brown, ed.) (Pergamon Press, Oxford, 1967), pp. 907–914

4 Kaminetzky, L., and Keller, J.B.: 'Diffraction coefficients for higher order edges and vertices', *SIAM J. Appl. Math.*, Vol. 22, pp. 109–134, 1972

5 Miller, M.A., and Talanov, V.I.: 'Electromagnetic surface waves guided by small-curvature boundaries', *Zh. Tekh. Fiz.*, Vol. 26, pp. 2755–2765, 1956 = *Sov. Phys. Tech. Phys.*, Vol. 1, pp. 2665–2673, 1956

6 Buslaev, V.S.: 'On the asymptotic behavior of the spectral characteristics of exterior problems for the Schrödinger operator', *Izv. Akad. Nauk SSSR, Ser. Mat.*, Vol. 39, pp. 149–235, 1975 = *Math. USSR Izvestija*, Vol. 9, pp. 139–223, 1975

7 Rulf, B.: 'Relation between creeping waves and lateral waves on a curved interface', *J. Math. Phys.*, Vol. 8, pp. 1785–1793, 1967

8 Wait, J.R. 'On surface wave propagation around a bend', *Acta Phys. Austriaca*, Vol. 32, pp. 113–121, 1970

9 Babich, V.M.: 'High-frequency oscillations of a point source near a concave mirror', *Zapiski Nauchn. Sem. Leningrad Otd. Mat. Inst. ANSSSR*, Vol. 51, pp. 5–19, 1975

10 Frisk, G.V., and Überall, H.: 'Creeping waves and lateral waves in acoustic scattering by large elastic cylinders', *J. Acoust. Soc. Amer.*, Vol. 59, pp. 46–54, 1976

11 Felsen, L.B., and Green, A.: 'Excitation of large concave surfaces', IEEE AP-S International Symposium Digest, 1976, pp. 374–377

Appendix A

Summation of series

A.1 Riemann's zeta function

In Chapters 3 and 5 we encounter certain sums involving inverse powers of numbers. By various manipulations, described below, these sums can be put in terms of the sums of the inverse powers of the natural numbers which are given by Riemann's ζ-function, defined as

$$\zeta(r) = \sum_{1}^{\infty} 1/n^r \tag{A1}$$

When r is even the summation can be expressed in closed form involving powers of π; in particular we have

$$\sum_{1}^{\infty} 1/n^2 = \pi^2/6, \qquad \sum_{1}^{\infty} 1/n^4 = \pi^4/90 \tag{A2}$$

When r is odd, a comparable formula does not exist and we have, to four decimal places, the numerical values

$$\sum_{1}^{\infty} 1/n^3 = \zeta(3) = 1{\cdot}2021, \qquad \sum_{1}^{\infty} 1/n^5 = \zeta(5) = 1{\cdot}0369 \tag{A3}$$

Many of the sums encountered are summed over odd integers only. By extracting the even powers we get

$$\sum_{1}^{\infty} 1/n^2 = \sum_{\substack{1\\ \text{odd}}}^{\infty} 1/n^2 + \sum_{\substack{2\\ \text{even}}}^{\infty} 1/n^2 = \sum_{\substack{1\\ \text{odd}}}^{\infty} 1/n^r + 2^{-r} \sum_{1}^{\infty} 1/n^r$$

Hence

$$\sum_{\substack{1\\ \text{odd}}}^{\infty} 1/n^2 = (1 - 2^{-r}) \sum_{1}^{\infty} 1/n^2 \tag{A4}$$

In particular

$$\sum_{\substack{1\\ \text{odd}}}^{\infty} 1/n^2 = \pi^2/8, \qquad \sum_{\substack{1\\ \text{odd}}}^{\infty} 1/n^4 = \pi^4/96 \tag{A5}$$

A.2 Some general expressions

Let $f(x)$ be an odd function of its argument, and consider the sum

$$S_0 = \sum_{n=0}^{\infty}{}' \left[f(n-m) - f(n+m)\right] \tag{A6}$$

where *either* m is an even integer and the summation is taken over odd values of n, *or* m is an odd integer and the summation is taken over even values of n. In either case $n \pm m$ will be odd. The prime on the summation sign is used to indicate that n does not assume all integral values.

Writing $-n$ for n in the first summation and making use of the property that f is an odd function, gives

$$S_0 = -\sum_{0}^{\infty}{}' f(n+m) - \sum_{0}^{-\infty}{}' f(n-m).$$

When m is even so that n covers only odd integers the $n = 0$ term is absent and the formula gives

$$S_0 = -\sum_{\substack{-\infty \\ n\,\mathrm{odd}}}^{\infty} f(n+m) = \sum_{\substack{-\infty \\ n'\,\mathrm{odd}}}^{\infty} f(n') = 0 \tag{A7}$$

since $f(n')$ is an odd function.

When m is odd and n covers the even integers an extra term enters at $n = 0$ and the same method gives

$$S_0 = -f(m) + \sum_{\substack{-\infty \\ n\,\mathrm{even}}}^{\infty} f(n+m) = -f(m) + \sum_{\substack{-\infty \\ n'\,\mathrm{odd}}}^{\infty} f(n') = -f(m) \tag{A8}$$

Similar considerations apply for an even function $g(x)$. Then

$$S_e = \sum_{0}^{\infty}{}' g(n-m) + \sum_{0}^{\infty} g(n+m) = \sum_{0}^{-\infty}{}' g(n+m) + \sum_{0}^{\infty}{}' g(n+m).$$

When m is even, so n covers the odd integers, the $n = 0$ term is absent and

$$S_e = \sum_{\substack{-\infty \\ n\,\mathrm{odd}}}^{\infty} g(n+m) = \sum_{\substack{-\infty \\ n'\,\mathrm{odd}}}^{\infty} g(n') = 2 \sum_{n\,\mathrm{odd}} g(n) \tag{A9}$$

Similarly, when m is odd and n covers the even integers, there is an extra term at $n = 0$, and

$$S_e = g(m) + \sum_{\substack{-\infty \\ n\,\mathrm{even}}}^{\infty} g(n+m) = g(m) + \sum_{\substack{-\infty \\ n'\,\mathrm{odd}}}^{\infty} g(n') = g(m) + 2 \sum_{\substack{1 \\ n\,\mathrm{odd}}}^{\infty} g(n) \tag{A10}$$

An 'interchange' formula, useful for even functions $g(x)$ and $h(x)$, occurs when m is even and n covers the odd integers.

$$S = \sum_{\substack{1 \\ n\text{ odd}}}^{\infty} h(n)[g(n+m)+g(n-m)] = \sum_{\substack{1 \\ n\text{ odd}}}^{\infty} h(n)g(n+m)$$

$$+ \sum_{\substack{-1 \\ n\text{ odd}}}^{-\infty} h(-n)g(n+m) = \sum_{\substack{-\infty \\ n\text{ odd}}}^{\infty} h(n)g(n+m)$$

$$= \sum_{\substack{-\infty \\ n'\text{odd}}}^{\infty} h(n'-m)g(n') = \sum_{\substack{1 \\ n\text{ odd}}}^{\infty} g(n)[h(n+m)+h(n-m)] \tag{A11}$$

This result was utilized in the calculations of section 5.9 to shift the 'dummy' index m from $\beta_{n\pm m}$ to other factors in the equations.

A.3 A summation over even integers

As an example consider the sum $S = \sum_{\substack{2 \\ n\text{ even}}}^{\infty} (2n^2+1)/(n^2-1)^4$ of section 3.6. To use the formulas of section A.2 we take the following relations in order to put $(2n^2+1)/(n^2-1)^4$ into a suitable partial fraction form,

$$\begin{aligned}
\frac{1}{(n+1)^4} + \frac{1}{(n-1)^4} &= \frac{2(n^4+6n^2+1)}{(n^2-1)^4} \\
\frac{1}{(n+1)^3} - \frac{1}{(n-1)^3} &= -\frac{2(3n^4-2n^2-1)}{(n^2-1)^4} \\
\frac{1}{(n+1)^2} + \frac{1}{(n-1)^2} &= \frac{2(n^6-n^4-n^2+1)}{(n^2-1)^4} \\
\frac{1}{n+1} - \frac{1}{n-1} &= -\frac{2(n^6-3n^4+3n^2-1)}{(n^2-1)^4}
\end{aligned} \tag{A12}$$

If we multiply these equations by α, β, γ and δ respectively, add, require the coefficients of n^6 and n^4 to vanish, the coefficient of n^2 to be 2, and that of n^0 to be 1, it is readily found that

$$\alpha = 3/16, \quad \beta = 1/8, \quad \gamma = \delta = 3/32 \tag{A13}$$

Hence

$$\sum_{\substack{2 \\ n\text{ even}}}^{\infty} \frac{(2n^2+1)}{(n^2-1)^4} = \sum_{\substack{2 \\ n\text{ even}}}^{\infty} \{\alpha[f_4(n+1)+f_4(n-1)] + \beta[f_3(n+1)$$

$$-f_3(n-1)] + \gamma[f_2(n+1)+f_2(n-1)] + \delta[f_1(n+1)-f_1(n-1)]\} \tag{A14}$$

with $f_r(x) = 1/x^r$.

Apart from the missing terms at $n = 0$ the summations are covered by equations (A8) and (A10), since clearly $f_r(x)$ is even when r is even, and odd when r is odd. Hence using (A2), and supplying the $n = 0$ terms,

$$\begin{aligned} S &= \alpha(\pi^4/48 - 1) + \beta(-1) + \gamma(\pi^2/4 - 1) + \delta(-1) \\ &= (\pi^4/16 + 3\pi^2/8 - 8)/16 \end{aligned} \tag{A15}$$

on replacing α, β, γ and δ by their values from (A13). This is the result quoted in the text.

A.4 A summation over odd integers

A typical summation occurring in the calculations of section 5.9 is

$$S = \sum_{\substack{1 \\ n\ \text{odd}}}^{\infty} \frac{1}{n^4}\left[\frac{1}{(n+m)^4} + \frac{1}{(n-m)^4}\right] \tag{A16}$$

with m an even integer.

In this form the expressions involving $n \pm m$ are already in separated form. In another instance the structure involves $2(m^2 + n^2)/(n^2 - m^2)^3$, and this must first be separated by putting it in the form

$$\begin{aligned} \frac{2(m^2+n^2)}{(n^2-m^2)^3} = &-\frac{1}{2m}\left[\frac{1}{(n+m)^3} - \frac{1}{(n-m)^3}\right] - \frac{1}{4m^2}\left[\frac{1}{(n+m)^2}\right. \\ &\left. + \frac{1}{(n-m)^2}\right] - \frac{1}{4m^3}\left[\frac{1}{n+m} - \frac{1}{n-m}\right] \end{aligned} \tag{A17}$$

Although the considerations of section A.2 apply to (A17) as it stands, the actual expressions encountered involve also inverse powers of n, as in (A16). In order to deal with this factor, a further partial fraction decomposition is needed. Let x and y be variables; then we have the decomposition

$$\frac{1}{x^2(x-y)} = \frac{1}{y^2(x-y)} - \frac{1}{x^2 y} - \frac{1}{y^2 x} \tag{A18}$$

Differentiating with respect to y successively gives

$$\frac{1}{x^2(x-y)^2} = \frac{1}{x^2y^2} + \frac{2}{xy^3} - \frac{2}{y^3(x-y)} + \frac{1}{y^2(x-y)^2} \tag{A19}$$

$$\frac{1}{x^2(x-y)^3} = -\frac{1}{x^2y^3} - \frac{3}{xy^4} + \frac{3}{y^4(x-y)} - \frac{2}{y^3(x-y)^2} + \frac{1}{y^2(x-y)^3} \tag{A20}$$

$$\frac{1}{x^2(x-y)^4} = \frac{1}{y^4x^2} + \frac{4}{y^5x} - \frac{4}{y^5(x-y)} + \frac{3}{y^4(x-y)^2} - \frac{2}{y^3(x-y)^3} + \frac{1}{y^2(x-y)^4} \tag{A21}$$

In these expressions take $x = n$ (odd) and $y = m$ (even) and apply (A8), (A10) and (A2). The following results, needed in section 3, are found.

$$\sum_1^{\infty}{}' \frac{1}{n^2}\left[\frac{1}{n+m} - \frac{1}{n-m}\right] = \pi^2/4m \tag{A22}$$

$$\sum_1^{\infty}{}' \frac{1}{n^2}\left[\frac{1}{(n+m)^2} - \frac{1}{(n-m)^2}\right] = \pi^2/2m^2 \tag{A23}$$

$$\sum_1^{\infty}{}' \frac{1}{n^2}\left[\frac{1}{(n+m)^3} - \frac{1}{(n-m)^3}\right] = 3\pi^3/4m^3 \tag{A24}$$

$$\sum_1^{\infty}{}' \frac{1}{n^2}\left[\frac{1}{(n+m)^4} + \frac{1}{(n-m)^4}\right] = \pi^2/m^4 + \pi^4/48m^2 \tag{A25}$$

(The prime on the summation is to act as a reminder that n is summed over the odd integers only.)

In a similar way, starting with the partial fraction decomposition of $1/x^4(x-y)$ and differentiating successively with respect to y gives

$$\frac{1}{x^4(x-y)} = \frac{-1}{x^4y} - \frac{1}{x^3y^2} - \frac{1}{x^2y^3} - \frac{1}{xy^4} + \frac{1}{y^4(x-y)} \tag{A26}$$

$$\frac{1}{x^4(x-y)^2} = \frac{1}{x^4y^2} + \frac{2}{x^3y^3} + \frac{3}{x^2y^4} + \frac{4}{xy^5} - \frac{4}{y^5(x-y)} + \frac{1}{y^4(x-y)^2} \tag{A27}$$

$$\frac{1}{x^4(x-y)^3} = \frac{-1}{x^4y^3} - \frac{3}{x^3y^4} - \frac{6}{x^2y^5} - \frac{10}{xy^6} + \frac{10}{y^6(x-y)} - \frac{4}{y^5(x-y)^2} + \frac{1}{y^4(x-y)^3} \tag{A28}$$

$$\frac{1}{x^4(x-y)^4} = \frac{1}{x^4y^4} + \frac{4}{x^3y^5} + \frac{10}{x^2y^6} + \frac{20}{xy^7} - \frac{20}{y^7(x-y)} + \frac{10}{y^6(x-y)^2} - \frac{4}{y^5(x-y)^3} + \frac{1}{y^4(x-y)^4} \tag{A29}$$

Taking $x = n, y = m$ gives

$$\sum_1^{\infty}{}' \frac{1}{n^4}\left[\frac{1}{n+m} - \frac{1}{n-m}\right] = \pi^4/48m + \pi^2/4m^3 \qquad \text{(A30)}$$

$$\sum_1^{\infty}{}' \frac{1}{n^4}\left[\frac{1}{(n+m)^2} + \frac{1}{(n-m)^2}\right] = \pi^4/48m^2 + \pi^2/m^4 \qquad \text{(A31)}$$

$$\sum_1^{\infty}{}' \frac{1}{n^4}\left[\frac{1}{(n+m)^3} - \frac{1}{(n-m)^3}\right] = \pi^4/48m^3 + 5\pi^2/2m^5 \qquad \text{(A32)}$$

$$\sum_1^{\infty}{}' \frac{1}{n^4}\left[\frac{1}{(n+m)^4} + \frac{1}{(n-m)^4}\right] = \pi^4/24m^4 + 5\pi^2/m^6 \qquad \text{(A33)}$$

The last expression gives the result for (A16). In all these formulae it should be recalled that m is an even integer and that n is summed over the odd integers only.

With the help of these results the summations involved in the calculations of section 5.9 can be simplified, leading eventually to (5.51) for the even-mode coupling coefficients.

Appendix B

Mode product integrals

B.1 Integration methods

The integrations involved in (5.50) all involve, one way or another, evaluations over the cross-section of products of polynomials and trigonometric functions. They are therefore basically elementary, and the main point in discussing their integration here is to indicate methods of avoiding long and tedious operations of multiple integration-by-parts.

By way of typical examples we shall here evaluate

$$I_1 = \int \phi_n(\eta)\xi_m(\eta)d\eta$$

$$I_2 = \int \phi_n(\eta)\xi_0^2(\eta)d\eta$$

$$I_3 = \int \phi_n(\eta)\zeta_0(\eta)d\eta$$

$$I_4 = \int \xi_0(\eta)\xi_m(\eta)d\eta$$

In all cases the integration is from $-b/2$ to $b/2$ and the limits will not be shown explicitly.

For integrals involving ϕ_n and ξ_0 or ζ_0 the method is to use (4.14) to replace ϕ_n in the form

$$\phi_n = -(b^2/n^2\pi^2)\phi_n'' \tag{B1}$$

and then to integrate by parts twice, noting that ϕ_n', ξ_m' and ζ_0' all vanish at the limits. Since ξ_0 is a cubic and ζ_0 is sextic, a repetition of this process up to three times is sufficient to reduce the integrand to ϕ_n times a constant; the latter, of course, then integrates to zero. Integrations involving ξ_m can be achieved through the use of (4.28), together with the cosine product relation which gives $2\phi_n\phi_m = \phi_{n+m} + \phi_{n-m}$. Since ξ_0 is purely a polynomial whilst ξ_m involves also ϕ_m and ϕ'_m, the case $m = 0$ is special, and is not included under I_1 and I_4 above. The integrals may be found readily by the same methods, however. In the following all expressions are taken between the limits $\pm b/2$.

B.2 Evaluation of I_1

From (4.28) with $B_{m1} = 0$ we have

$$\xi''_m + \frac{m^2\pi^2}{b^2}\xi_m + 2\eta\phi'' + \phi'_m + 2\eta k^2\phi_m = 0 \tag{B2}$$

Multiply this equation by ϕ_n and integrate twice by parts the term in ξ''_m.

$$\frac{m^2\pi^2}{b^2} I_1 + \left[\xi'_m\phi_n - \xi_m\phi'_n + \int \xi_m\phi''_n\, d\eta\right]$$
$$+ \int \phi_n[2\eta\phi''_m + \phi'_m + 2\eta k^2\phi_m]\, d\eta = 0 \tag{B3}$$

Recalling that $\phi''_n = -(n^2\pi^2/b^2)\phi_n$ and that ϕ'_n and ξ'_m are zero at the limits the equation can be written, on further using (B1) and the cosine product relation,

$$I_1\pi^2(n^2 - m^2)/b^2 = (m^2 - k^2b^2/\pi^2)\int \eta\left[\frac{\phi''_{n+m}}{(n+m)^2} + \frac{\phi''_{n-m}}{(n-m)^2}\right] d\eta$$
$$+\frac{m}{2}\int\left(\frac{\phi'_{m+n}}{m+n} + \frac{\phi'_{m-n}}{m-n}\right) d\eta \tag{B4}$$

The integrations are now quite straightforward and give

$$I_1 = (2b^2/\pi^2)[2(m^2 + n^2)k^2b^2/\pi^2 - m^2(m^2 + 3n^2)]/(n^2 - m^2)^3 \tag{B5}$$

B.3 Evaluation of I_2

On using (B1) and integrating by parts successively we get

$$I_2 = \int \phi_n\xi_0^2\, d\eta = -(b^2/n^2\pi^2)\int \phi''_n\xi_0^2 d\eta$$
$$= -(b/n\pi)^2\left[\phi'_n\xi_0^2 - 2\phi_n\xi_0\xi'_0 + 2\int \phi_n(\xi_0\xi''_0 + \xi'^2_0)d\eta\right]$$
$$= 2(b/n\pi)^4\left[\phi'_n(\xi_0\xi''_0 + \xi'^2_0) - \phi_n(\xi_0\xi'''_0 + 3\xi'_0\xi''_0)\right.$$
$$\left. + \int \phi_n(4\xi'_0\xi'''_0 + 3\xi''^2_0)d\eta\right] \tag{B6}$$

Since ξ_0 is a cubic $\xi_0^{iv} = 0$. From (4.5) $\xi_0''' = -2k^2$, and from (4.7), $\xi_0(b/2) = k^2 b^3/12$. Noting that n is even (otherwise $I_2 = 0$), (B6) gives

$$I_2 = 2(b/n\pi)^4 \left[k^4 b^3/3 + \int \phi_n(4\xi_0' \xi_0''' + 3\xi_0''^2)d\eta \right] \tag{B7}$$

Both functions of ξ_0 are quadratic so two more integrations by parts suffice.

$$\begin{aligned} I_2 &= 2(b/n\pi)^4 \{k^4 b^3/3 - (b/n\pi)^2 [\phi_n'(4\xi_0'\xi_0''' + 3\xi_0''^2) - \phi_n(10\xi_0''\xi_0''')]\} \\ &= \frac{2k^4 b^7}{3n^2\pi^2}\left(1 + \frac{120}{n^2\pi^2}\right), \quad (n \text{ even}) \end{aligned} \tag{B8}$$

on evaluation at the limits.

B.4 Evaluation of I_3

Since $\zeta_0(\eta)$ from (4.11) is of sixth degree, a multiple integration by parts six times is necessary to reduce the integral. Using (B1) we get

$$I_3 = -(b/n\pi)^2 \int \phi_n''\zeta_0 d\eta = -(b/n\pi)^2 \left[\phi_n'\zeta_0 - \phi_n\zeta_0' + \int \phi_n\zeta_0'' d\eta \right], \tag{B9}$$

The integrated part vanishes at the limits. From (4.8),

$$\xi_0'' = k^2(4\eta^2 + B_2 - b^2/4) - 2k^2\eta\xi_0 \tag{B10}$$

continuing the process gives

$$I_3 = (b/n\pi)^4 \left[\phi_n'\zeta_0'' - \phi_n\zeta_0''' + \int \phi_n\zeta_0^{iv} d\eta \right] \tag{B11}$$

Now $\xi_0''' = 8k^2\eta - 2k^2\xi_0 - 2k^2\xi_0'$, which equals $k^2 b(4 - k^2b^2/6)$ at $\eta = b/2$. Recalling that ζ_0 is even so that $I_3 = 0$ unless n is also even (B11) gives

$$I_3 = (b/n\pi)^4 [-k^2 b(8 - k^2b^2/3) - (b/n\pi)^2(\phi_n'\zeta^{iv} - \phi_n\zeta^{v})] \tag{B12}$$

From (4.11), $\zeta^v = 16k^2\eta = 8k^4 b$ at $\eta = b/2$. Hence (B12) becomes

$$I_3 = (b/n\pi)^4 [-k^2 b(8 - k^2b^2/3) + (b/n\pi)^2 16k^4 b], \quad n \text{ even} \tag{B13}$$

B.5 Evaluation of I_4

As already remarked, ξ_m is of a special character when $m = 0$, and the present calculation avoids this case, though the methods of this appendix still apply. Moreover, since ξ_m is odd or even according as m is even or odd, and ξ_0 is odd, I_4 will be zero unless m is even, and the present calculation assumes this.

We start with (B2), multiply by ξ_0 and integrate by parts twice on the first term.

$$0 = \xi_m' \xi_0 - \xi_m \xi_0' + \int \xi_m \xi_0'' d\eta + (m^2\pi^2/b^2) I_4$$
$$+ \int \phi_m' \xi_0 d\eta + 2\beta_m^2 \int \eta \phi_m \xi_0 d\eta \tag{B14}$$

Again, from (B2), multiply by $\xi_0' = -2yk^2$ and integrate,

$$0 = \xi_m' \xi_0'' - \xi_m \xi_0''' + (m^2\pi^2/b^2) \int \xi_m \xi_0'' d\eta$$
$$+ \int \phi_m' \xi_0'' d\eta + 2\beta_m^2 \int \eta \phi_m \xi_0'' d\eta \tag{B15}$$

Now multiply the second equation by $(b/m\pi)^2$ and subtract from the first, use (4.18) to evaluate ξ_m at $\eta = b/2$, and integrate the ϕ_m' term by parts. The result of these operations can be put in the form

$$0 = (m\pi/b)^2 I_4 + b(kb/m\pi)^4 + \phi_m(\xi_0 - \xi_0'' b^2/m^2\pi^2) + J \tag{B16}$$

where $$J = \int \phi_m [2\eta\beta_m^2(\xi_0 - \xi_0'' b^2/m^2\pi^2) - \xi_0'] d\eta$$

Since the polynomial in J is of the fourth degree, four more integrations by parts are needed. Dropping the integrated parts involving ϕ_m', which vanishes at the limits, gives

$$J = (b^2\phi_m/m^2\pi^2)[2\beta_m^2(\xi_0 - \xi_0'' b^2/m^2\pi^2 + \eta\xi_0' - \eta\xi_0''' b^2/m^2\pi^2)$$
$$- \xi_0'' - 2\beta_m^2(3\xi_0'' + \eta\xi_0''')b^2/m^2\pi^2] \tag{B17}$$

This has to be evaluated at the limits and the result inserted into (B16). After some re-arrangement we get

$$I_4 = \frac{k^2 b^5}{6m^2\pi^2}\left[1 + \frac{120 - 2k^2b^2}{m^2\pi^2} - \frac{150k^2b^2}{m^4\pi^4}\right], \quad (m \text{ even}) \tag{B18}$$

Appendix C

Some asymptotic expressions for ratios of cylinder functions

C.1 Formulas from Debye's expansions

In Chapters 7 and 10, we have need of the asymptotic forms of certain combinations of cylinder functions (or their derivatives) of the form

$$\frac{Z_\nu(\nu z + bz)}{Z_\nu(\nu z)}$$

when ν is large, and b and z are fixed numbers on the order of unity. In this appendix, we will give a method for obtaining these forms from the known (Debye) asymptotic forms of Z_ν.

Lemma: Let ν be large, z and b finite, and let a function $F_\nu(\nu z)$ possess the asymptotic expansion, for large ν, of the form

$$F_\nu(\nu z) = \exp\,[\nu f(z) + g(z)]\,\{1 + h(z)/\nu + 0(1/\nu^2)\} \qquad \text{(C1)}$$

Then using Taylor's expansion we get

$$\frac{F_\nu(\nu z + bz)}{F_\nu(\nu z)} = \exp\,[bzf'(z)]\left\{1 + \frac{bzg'(z) + b^2z^2f''(z)/2}{\nu} + 0(1/\nu^2)\right\} \qquad \text{(C2)}$$

Applying this lemma to $F_\nu(\nu z) = -(2\nu/\pi)^{1/2} K_\nu'(\nu z)$ and using Debye's expansion[1] for which

$$f(z) = \log\frac{1 + (1 + z^2)^{1/2}}{z} - (1 + z^2)^{1/2}$$

$$g(z) = (1/4)\log(1 + z^2) - \log z$$

we have the following result:

$$\frac{K_\nu'(\nu z + bz)}{K_\nu'(\nu z)} = \exp\left[-b(1+z^2)^{1/2}\right]\left\{1 + \frac{b}{2\nu}\left[-\frac{2+z^2}{1+z^2} + \frac{b}{(1+z^2)^{1/2}}\right] + 0(\nu^{-2})\right\} \tag{C3}$$

In a similar way, we get

$$\frac{K_\nu(\nu z + bz)}{K_\nu(\nu z)} = \exp\left[-b(1+z^2)^{1/2}\right]\left\{1 + \frac{b}{2\nu}\left[-\frac{z^2}{1+z^2} + \frac{b}{(1+z^2)^{1/2}}\right] + 0(\nu^{-2})\right\} \tag{C4}$$

$$\frac{J_\nu'(\nu y + by)}{J_\nu'(\nu y)} = \exp\left[b(1-y^2)^{1/2}\right.\left\{1 + \frac{b}{2\nu}\left[-\frac{2-y^2}{1-y^2} - \frac{b}{(1-y^2)^{1/2}}\right] + 0(\nu^{-2})\right\} \tag{C5}$$

$$\frac{Y_\nu'(\nu y + by)}{Y_\nu'(\nu y)} = \exp\left[-b(1-y^2)^{1/2}\right]\left\{1 + \frac{b}{2\nu}\left[-\frac{2-y^2}{1-y^2} + \frac{b}{(1-y^2)^{1/2}}\right] + 0(\nu^{-2})\right\} \tag{C6}$$

$$\frac{J_\nu(\nu y + by)}{J_\nu(\nu y)} = \exp\left[b(1-y^2)^{1/2}\right]\left\{1 + \frac{b}{2\nu}\left[\frac{y^2}{1-y^2} - \frac{b}{(1-y^2)^{1/2}}\right] + 0(\nu^{-2})\right\} \tag{C7}$$

$$\frac{Y_\nu(\nu y + by)}{Y_\nu(\nu y)} = \exp\left[-b(1-y^2)^{1/2}\right]\left\{1 + \frac{b}{2\nu}\left[\frac{y^2}{1-y^2} + \frac{b}{(1-y^2)^{1/2}}\right] + 0(\nu^{-2})\right\} \tag{C8}$$

Expressions (C5) through (C8) carry the additional restriction $|y| < 1$.

In addition, from Debye's formulas,

$$J_\nu'(\nu y)/Y_\nu'(\nu y) \sim \tfrac{1}{2}\exp\left[-2\nu\bar{f}(y)\right] \tag{C9}$$

$$J_\nu(\nu y)/Y_\nu(\nu y) \sim -\tfrac{1}{2}\exp\left[-2\nu\bar{f}(y)\right] \tag{C10}$$

where

$$\bar{f}(y) = \tanh^{-1}(1-y^2)^{1/2} - (1-y^2)^{1/2} \tag{C11}$$

and so from (C5) – (C10) we obtain

$$\frac{H_\nu^{(2)\prime}(\nu y + by)}{H_\nu^{(2)\prime}(\nu y)} \sim \exp\left[-b(1-y^2)^{1/2}\right]\left\{1 + \frac{b}{2\nu}\left[-\frac{2-y^2}{1-y^2} + \frac{b}{(1-y^2)^{1/2}}\right]\right\} + \frac{j}{2}\exp\left[-2\nu\bar{f}(y) + b(1-y^2)^{1/2}\right] \tag{C12}$$

$$\frac{H_\nu^{(2)}(\nu y + by)}{H_\nu^{(2)}(\nu y)} \sim \exp\left[-b(1-y^2)^{1/2}\right]\left\{1 + \frac{b}{2\nu}\left[\frac{y^2}{1-y^2} + \frac{b}{(1-y^2)^{1/2}}\right]\right\} - \frac{j}{2}\exp\left[-2\nu\bar{f}(y) + b(1-y^2)^{1/2}\right] \tag{C13}$$

C.2 Limiting form for equal arguments

Some further results follow as an extension of the lemma:

Corollary: Let $F_\nu(\nu z)$ be as in the lemma. By forming

$$\frac{1}{bz}\left[\frac{F_\nu(\nu z + bz)}{F_\nu(\nu z)} - 1\right]$$

and taking the limit as b (not z) goes to zero, we have

$$\frac{F_\nu'(\nu z)}{F_\nu(\nu z)} = f'(z) + \frac{g'(z)}{\nu} + 0(\nu^{-2}) \tag{C14}$$

From this corollary we obtain

$$\frac{J_\nu'(\nu y)}{J_\nu(\nu y)} = \frac{(1-y^2)^{1/2}}{y} + \frac{1}{2\nu}\left[\frac{y}{1-y^2}\right] + 0(\nu^{-2}) \tag{C15}$$

$$\frac{Y_\nu'(\nu y)}{Y_\nu(\nu y)} = -\frac{(1-y^2)^{1/2}}{y} + \frac{1}{2\nu}\left[\frac{y}{1-y^2}\right] + 0(\nu^{-2}) \tag{C16}$$

and thus from (C9) and (C10)

$$\frac{H_\nu^{(2)\prime}(\nu y)}{H_\nu^{(2)}(\nu y)} \sim \left\{-\frac{(1-y^2)^{1/2}}{y} + \frac{1}{2\nu}\left[\frac{y}{1-y^2}\right]\right\} \cdot \left\{1 + j\exp\left[-2\nu\bar{f}(y)\right]\right\} \tag{C17}$$

Reference

1 Abramowitz, M., and Stegum, I.A. (ed.): 'Handbook of mathematical functions' (Applied Mathematics Series 55) Washington, D.C., NBS, June 1964, pp. 366 and 378

Appendix D

Determination of the reflection coefficient

D.1 Asymptotic method and Stokes lines

In this appendix, an explicit expression for the reflection coefficient $\mathcal{R}$ is determined. Following an asymptotic procedure adopted by Budden[1] in what he called Furry's method, we first observe that the expression

$$\zeta^{-1/4}\,[Ae^{-(2/3)\zeta^{3/2}} + Be^{(2/3)\zeta^{3/2}}] \tag{D1}$$

where A, B are two arbitrary constants, forms an asymptotic solution of Stokes' differential equation

$$\left(\frac{d^2}{d\zeta^2} + \zeta\right)\Phi(\zeta) = 0 \tag{D2}$$

for large ζ. Exact solutions of this equation as combinations of Airy functions $Ai(\zeta)$ and $Bi(\zeta)$ are known to be single-valued in ζ. Since the expression in (D1) is multi-valued, it is clear that the solution of (D2) cannot be represented by the same combination of A, B for all ζ. This observation is of course consistent with our adoption of two different expressions in (7.52) and (7.54) for the region where ζ is positive real and the region where ζ is negative real.

In order to find the various combinations of A and B, it is convenient to divide the complex ζ-plane into different regions according to the different behaviors the two terms in (D1) exhibit in these regions. We note that the exponents of both exponential terms are real if

$$\arg\zeta = 0, \quad 2\pi/3 \quad \text{and} \quad 4\pi/3 \tag{D3}$$

Thus, as $|\zeta| \to \infty$, one of the exponentials becomes infinitely larger than the other one. Typically, the exponentially-increasing term is referred to as the dominant term, the exponentially-decreasing term as the sub-dominant term, and the radial lines in (D3) as Stokes lines.

Now since we are normally only interested in the leading term in an asymptotic expansion, an additional sub-dominant term can in principle be inserted into the asymptotic expression anytime we cross a Stokes line and the error thus introduced will be negligible in the asymptotic sense. However the amplitude of this additional term has to be proportional to the amplitude of the dominant term so that no error will be incurred in the event the dominant terms is absent in the asymptotic expression.

By analogy, we can also define anti-Stokes lines when the two exponents in (D1) have equal moduli. From (D1) it is clear that they are located at

$$\arg \zeta = \pi/3, \quad \pi \quad \text{or} \quad 5\pi/3 \tag{D4}$$

Thus, as $|\zeta|$ increases along the radial lines, both terms remain purely oscillatory with an increasing rate of oscillation. Obviously, the roles of dominant and sub-dominant terms are interchanged whenever we cross an anti-Stokes line. The situation is illustrated in Fig. D.1 where the Stokes lines are denoted by solid lines S_1, S_2 and S_3 and the anti-Stokes lines are denoted by dashed lines A_1, A_2 and A_3.

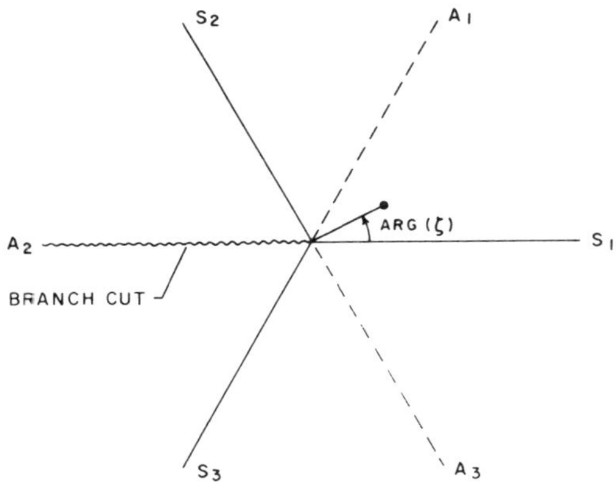

Fig. D.1 Stokes and anti-Stokes lines in the ζ-plane

In addition to the Stokes lines and anti-Stokes lines, we also define a branch cut along the negative real axis in the ζ-plane so that $\zeta^{1/2}$ equals $+j|\zeta|^{1/2}$ in the region between S_2 and A_2 just above A_2. As it crosses S_2, (D1) is modified by adding a sub-dominant term $\lambda_1 A \exp(\frac{2}{3}\zeta^{3/2})$ so that we have

$$\zeta^{-1/4}\{Ae^{-(2/3)\zeta^{3/2}} + (B + \lambda_1 A)e^{+(2/3)\zeta^{3/2}}\} \tag{D5}$$

in the region between S_2 and A_2. Here, λ_1 is a constant as yet undetermined. Following the same procedure, it is not difficult to see that the corresponding expression in the region between A_2 and S_3, after the crossing at S_1 and S_3, is given by

$$\zeta^{-1/4}(A+\lambda_2 B)e^{-(2/3)\zeta^{3/2}} + [B+\lambda_3(A+\lambda_2 B)]\, e^{+(2/3)\zeta^{3/2}} \quad \text{(D6)}$$

Now, on crossing the branch cut from the top to bottom, $\zeta^{1/4}$ changes by a factor of $\exp(-j\pi/2)$ and $\zeta^{3/2}$ changes by $\exp(-j3\pi)$. A comparison of like terms in (D5) and (D6) then yields

$$A = j[B+\lambda_3(A+\lambda_2 B)]; \quad B+\lambda_1 A = j(A+\lambda_2 B)$$

for any A and B. Thus, we have

$$1 = j\lambda_3, \quad 1+\lambda_2\lambda_3 = 0, \quad \lambda_1 = j \quad \text{and} \quad 1 = j\lambda_2.$$

Or, more explicitly,

$$\lambda_2 = \lambda_3 = -j \quad \text{and} \quad \lambda_1 = j.$$

D.2 Solution matching

To apply the result just derived, we first set $A = 1$ and $B = 0$ so that the solution is explicitly given as

$$\Phi_1 \simeq \zeta^{-1/4}[e^{-(2/3)\zeta^{3/2}} + je^{(2/3)\zeta^{3/2}}]; \text{ for } \zeta \text{ negative and real } (\arg\zeta = \pi) \quad \text{(D7)}$$

and

$$\Phi_1 \simeq \zeta^{-1/4}e^{-(2/3)\zeta^{3/2}}; \quad \text{for } \zeta \text{ positive and real } (\arg\zeta = 0) \quad \text{(D8)}$$

Equations (D7) and (D8) then represent the proper asymptotic expansions of the first Airy function $Ai(\zeta)$ for a negative and a positive ζ. On the other hand, if we set $A = j$ and $B = 2$, we have

$$\Phi_2 \simeq \zeta^{-1/4}[je^{-(2/3)\zeta^{3/2}} + e^{(2/3)\zeta^{3/2}}] \quad \text{(D9)}$$

when ζ is negative and real, i.e. $\arg\zeta = \pi$, and

$$\Phi_2 \simeq \zeta^{-1/4}[je^{-(2/3)\zeta^{3/2}} + 2e^{(2/3)\zeta^{3/2}}] \simeq 2\zeta^{-1/4}e^{(2/3)\zeta^{3/2}} \quad \text{(D10)}$$

when ζ is positive and real, i.e. $\arg\zeta = 0$. Equations (D9) and (D10) are then known as the asymptotic expansions of the second Airy-function $Bi(\zeta)$, with real negative and real positive arguments. We note that in (D10), we have omitted the term in $\exp(-\frac{2}{3}\zeta^{3/2})$ since it is the subdominant term in that region. Now, in order to represent an outgoing wave in the region when $\arg\zeta = \pi$, we need to combine the solutions

Φ_1 and Φ_2 so that

$$\Phi = (-j\Phi_1 + \Phi_2)/2 = \zeta^{-1/4} e^{+(2/3)\zeta^{3/2}}, \tag{D11}$$

Consequently, the same combination now yields a proper solution for $\arg \zeta = 0$ as

$$\Phi = (-j\Phi_1 + \Phi_2)/2 = \zeta^{-1/4}\left[\frac{-j}{2} e^{-(2/3)\zeta^{3/2}} + e^{(2/3)\zeta^{3/2}}\right] \tag{D12}$$

Comparing (D12) with (7.52) the reflection coefficient is now obtained as

$$\hat{\mathcal{R}} = -j/2 \quad \text{or} \quad \mathcal{R} = -(j/2)e^{-2k_0 Q(u_t)} \tag{D13}$$

which is the expression given in (7.53).

Reference

1 Budden, K.G.: 'Radio waves in the ionosphere' (Cambridge University Press, 1961)

Appendix E

Spectral distribution

E.1 Derivation from the wave equation

In order to find the spectral function of $f_m(\rho, \psi) = K_m\,[k_0(\gamma^2 - 1)^{1/2}\rho]\, e^{-jm\psi}$ we first note that $f_m(\rho, \psi)$ satisfies the wave equation

$$\left[\frac{\partial^2}{\partial y^2} + \frac{\partial^2}{\partial z^2} - (\gamma^2 - 1)\right] f_m(\rho, \psi) = 0 \tag{E1}$$

so that the spectral function $\tilde{f}_m$ as defined by (7.59) is

$$\tilde{f}_m(y; \xi) = f_{m0}(\xi) e^{-(\gamma^2+\xi^2-1)^{1/2} k_0 y} \tag{E2}$$

To determine the constant $f_{m0}(\xi)$, we proceed to evaluate the Fourier transform according to the definition given in (7.59)

$$f_{m0}(\xi) = \frac{k_0}{2\pi} e^{k_0(\gamma^2+\xi^2-1)^{1/2} y} \int_{-\infty}^{\infty} K_m\,[k_0(\gamma^2 - 1)^{1/2}\rho]\, e^{-jm\psi}\, e^{-jk_0\xi z}\, dz \tag{E3}$$

Now because (E3) is an identity for all y, the value of $f_{m0}(\xi)$ can be obtained by allowing y to approach infinity. Using the large-argument expansion of K_m, we have from (E3) the following expression

$$f_{m0}(\xi) = \lim_{y \to \infty} \left(\frac{k_0}{8\pi[\gamma^2 - 1]^{1/2}}\right)^{1/2} e^{k_0(\gamma^2+\xi^2-1)^{1/2} y} \times$$

$$\int_{-\infty}^{\infty} e^{-j(k_0\xi z + m \tan^{-1} z/y)} e^{-k_0(\gamma^2-1)^{1/2}(y^2+z^2)^{1/2}} \frac{dz}{(y^2 + z^2)^{1/4}} \tag{E4}$$

E.2 Steepest-descent evaluation

To evaluate the integral, we first change variables by putting $z = y \sinh \tau$

and $\theta_0 = \tan^{-1}\,[\xi/(\gamma^2-1)^{1/2}]$, to obtain

$$y^{1/2}\int_{-\infty}^{\infty} \cosh^{1/2}\tau\, e^{-k_0 y(\xi^2+\gamma^2-1)^{1/2}} \cosh(\tau - j\theta_0)\, e^{-jm\tan^{-1}(\sinh\tau)} d\tau$$

For large y, the main contribution of the integrand comes from the region near the saddle-point $\tau = j\theta_0$. Following a typical steepest-descent method, the leading term of the integral is then

$$y^{1/2}\cos^{1/2}\theta_0\, e^{m\tanh^{-1}(\sin\theta_0)}\, e^{-k_0(\xi^2+\gamma^2-1)^{1/2}y} \times$$

$$\int_{-\infty}^{\infty} e^{-[k_0(\xi^2+\gamma^2-1)^{1/2}/2](\tau-j\theta_0)^2}\, d\tau$$

$$= \left(\frac{2\pi}{k_0}\right)^{1/2} \frac{(\gamma^2-1)^{1/4}}{(\gamma^2+\xi^2-1)^{1/2}} \left(\frac{(\xi^2+\gamma^2-1)^{1/2}+\xi}{(\gamma^2-1)^{1/2}}\right)^m e^{-k_0(\xi^2+\gamma^2-1)^{1/2}y} \tag{E5}$$

Substitution of (E5) into (E4) finally yields

$$f_{m0}(\xi) = \frac{1}{2}\,\frac{1}{(\gamma^2+\xi^2-1)^{1/2}} \left(\frac{\xi+(\xi^2+\gamma^2-1)^{1/2}}{(\gamma^2-1)^{1/2}}\right)^m \tag{E6}$$

Thus, the spectral representation of $K_m[k_0(\gamma^2-1)^{1/2}\rho]\,e^{-jm\psi}$ is given by

$$K_m[k_0(\gamma^2-1)^{1/2}\rho]\,e^{-jm\psi} = \int_{-\infty}^{\infty} \frac{1}{2(\gamma^2+\xi^2-1)^{1/2}}$$

$$\left[\frac{\xi+(\xi^2+\gamma^2-1)^{1/2}}{(\gamma^2-1)^{1/2}}\right]^m e^{-(\gamma^2+\xi^2-1)^{1/2}k_0 y}\, e^{jk_0\xi z}\, d\xi \tag{E7}$$

Appendix F

Approximation to the secular equation

F.1 Approximation for $\epsilon_c \approx \epsilon$

Eliminating Λ from (9.14) and (9.15) gives the 'exact' equation

$$(y^2 F_m + x^2 \bar{M}_m)(\epsilon y^2 F_m + \epsilon_c x^2 M_m) = m^2 (y^2 + x^2)(\epsilon y^2 + \epsilon_c x^2) \tag{F1}$$

We seek a solution with $\epsilon_c \approx \epsilon$ and $M_m \approx \bar{M}_m$. Put $M_m + \bar{M}_m = 2y^2 N_m$ (this is the $N_m(y)$ of (9.20)) and $M_m - \bar{M}_m = 2\Delta M$. Also put $\epsilon_c/\epsilon = \theta(\approx 1)$ and write $y^2/x^2 = z$. Then (F1) becomes, on dropping the subscript m,

$$(zF + y^2 N - \Delta M)(zF + \theta y^2 N + \theta \Delta M) = m^2 (z + 1)(z + \theta) \tag{F2}$$

This can be re-written as

$$[zF + \tfrac{1}{2}y^2(1 + \theta)N + \tfrac{1}{2}y^2(1 - \theta)N - \Delta M] \times$$
$$[zF + \tfrac{1}{2}y^2(1 + \theta)N - \tfrac{1}{2}y^2(1 - \theta)N + \Delta M + (\theta - 1)\Delta M]$$
$$= m^2 [z + \tfrac{1}{2}(1 + \theta) + \tfrac{1}{2}(1 - \theta)] [z + \tfrac{1}{2}(1 + \theta) - \tfrac{1}{2}(1 - \theta)] \tag{F3}$$

On multiplying out this gives

$$[zF + \tfrac{1}{2}y^2(1 + \theta)N]^2 - m^2 [z + \tfrac{1}{2}(1 + \theta)]^2$$
$$= [\tfrac{1}{2}y^2(1 - \theta)N - \Delta M]^2 - m^2(1 - \theta)^2/4$$
$$+ (1 - \theta)\Delta M [zF + \tfrac{1}{2}y^2(1 + \theta)N + \tfrac{1}{2}y^2(1 - \theta)N - \Delta M] \tag{F4}$$

For $\epsilon \approx \epsilon_c$ all terms on the right of (F4) are at least of second order in small quantities. If they are neglected the equation reduces to

$$[zF + \tfrac{1}{2}y^2(1 + \theta)N]^2 = m^2 [z + \tfrac{1}{2}(1 + \theta)]^2 \tag{F5}$$

Returning to the original variables this becomes, on taking the square root,

$$F_m(x) + \tfrac{1}{2}x^2 N_m(y)(1 + \epsilon_c/\epsilon) = \pm m[1 + \tfrac{1}{2}x^2(1 + \epsilon_c/\epsilon)/y^2] \qquad \text{(F6)}$$

This is a slightly more accurate form of (9.19). The latter can be recovered by replacing $\frac{1}{2}(1 + \epsilon_c/\epsilon)$ by unity. By retaining this factor (9.21) is replaced by the slightly more accurate

$$\frac{xJ_m(x)}{J_{m-1}(x)} = \frac{2\epsilon}{\epsilon + \epsilon_c} \cdot \frac{yK_m(y)}{K_{m-1}(y)} \left[1 + \frac{\sigma\epsilon_m}{2yK_m(y)K_{m-1}(y)}\right] \qquad \text{(F7)}$$

F.2 Polarization effects

The right hand side of (F4) simplifies a little to

$$(N^2y^4 - m^2)(1 - \theta)^2/4 + \theta(\Delta M)^2 + zF(1 - \theta)\Delta M \qquad \text{(F8)}$$

where

$$\Delta M = (\delta_{m\alpha} - \delta_{m\beta})/2K_m^2 = \sigma\epsilon_m e^{2im\alpha}/2K_m^2 \qquad \text{(F9)}$$

Now although all quantities in (F8) are of second order when $(1 - \theta)$ and ΔM are small, the last term is first order in ΔM, which is polarization sensitive according to (F9). Hence, when $1 - \theta$ is no longer small, the conclusion that the bend radiation is independent of the polarization relative to the plane of the bend ceases to hold. In fact, self-consistency of the phases of (9.14) and (9.15) further requires

$$\epsilon_c(y^2F + x^2M)\sin 2m\alpha + (\epsilon y^2F + \epsilon_c x^2M)\sin 2m\beta = 0 \qquad \text{(F10)}$$

where M is the value of M_m when $\delta_{m\alpha}$ is neglected. This equation is consistent with (9.17) only when $m\alpha = 0$ or $\pi/2$, which enables the right-hand side of (F9) to take the values $\pm\,\sigma\epsilon_m/2K_m^2$. [This quantity needs to be real anyhow, since (F1) has to give real values (in i) for x and y.] Thus the last term on the right-hand side of (F8) leads to the breaking of the polarization degeneracy, and the two polarizations propagate with slightly differing attenuations. An initial arbitrary polarization therefore has to be split up into two components relative to the plane of the bend. Only in the limit $\epsilon = \epsilon_c$ does (F10) become equivalent to (9.17) and the polarization-independent character ensue.

Appendix G

Angular two-dimensional divergence theorem

G.1 An application of Gauss's divergence theorem

In order to obtain the integral formula (9.31) from (9.30) we require a generalization of the two-dimensional divergence theorem which in Cartesian co-ordinates expresses the integral of the divergence of a vector function over a surface in terms of an integral over the boundary of that surface. Let us consider an arbitrary surface S bounded by a closed contour C which lies in a constant θ-plane as shown in Fig. G.1.

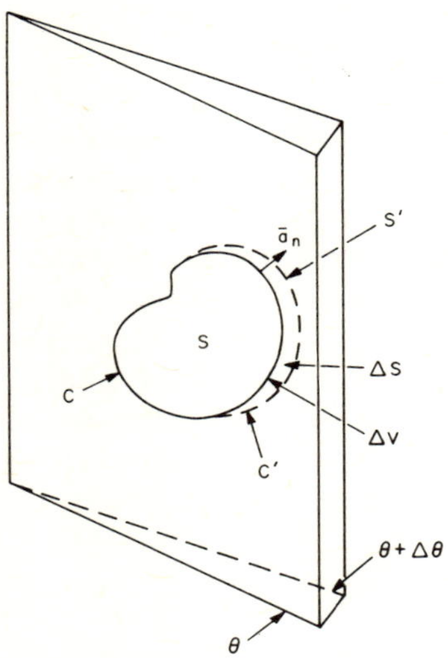

Fig. G.1 Diagram for proof of 2-dimensional divergence theorem transverse to θ

Now rotate this plane about the z-axis so that S moves to S′ in the plane $\theta + \Delta\theta$, tracing out the volume ΔV, and likewise C moves to C′ tracing out the lateral surface ΔS as shown. For any suitably differentiable vector function $\boldsymbol{F}$ on ΔV, applying the ordinary three-dimensional divergence theorem gives:

$$\int_{\Delta V} \operatorname{div} \boldsymbol{F} dV = \int_{S'} \boldsymbol{F}\cdot\boldsymbol{a}_\theta\, dS - \int_S \boldsymbol{F}\cdot\boldsymbol{a}_\theta\, dS + \int_{\Delta S} \boldsymbol{F}\cdot\boldsymbol{a}_n dS \tag{G1}$$

where $\boldsymbol{a}_n$ is the outward unit normal on ΔS. Dividing (G1) by $\Delta\theta$ and letting $\Delta\theta \to 0$, we have

$$\frac{1}{\Delta\theta}\int_{\Delta V} \operatorname{div} \boldsymbol{F} dV \to \int_S \rho \operatorname{div} \boldsymbol{F} dS$$

$$\frac{1}{\Delta\theta}\int_{\Delta S} \boldsymbol{F}\cdot\boldsymbol{a}_n dS \to \int_C \rho \boldsymbol{F}\cdot\boldsymbol{a}_n dl$$

which yields

$$\int_S \rho \operatorname{div} \boldsymbol{F} dS = \frac{d}{d\theta}\int_S \boldsymbol{F}\cdot\boldsymbol{a}_\theta\, dS + \oint_C \boldsymbol{F}\cdot\boldsymbol{a}_n dl \tag{G2}$$

or

$$\int_S \rho \operatorname{div}_t \boldsymbol{F} dS = \oint_C \rho \boldsymbol{F}\cdot\boldsymbol{a}_n dl \tag{G3}$$

which is the desired angular two-dimensional divergence theorem.

Appendix H

Phase correction

H.1 The reality of Δ/P

To demonstrate that (9.36) is purely a real quantity, we note that Maxwell's equations (9.30) for the straight guide can be expressed as a pair of coupled differential equations for E_{0s} and H_{0s} with real coefficients, and the transverse fields expressed as

$$\boldsymbol{E}_{0t} = -\frac{j}{k_0}\left\{\frac{\gamma_0}{n^2-\gamma_0^2}\nabla_t E_{0s} - \frac{1}{n^2-\gamma_0^2}\sqrt{\frac{\mu_0}{\epsilon_0}}\,\boldsymbol{a}_s \times \nabla_t H_{0s}\right\} \tag{H1}$$

$$\boldsymbol{H}_{0t} = -\frac{j}{k_0}\left\{\frac{\gamma_0}{n^2-\gamma_0^2}\nabla_t H_{0s} + \frac{n^2}{n^2-\gamma_0^2}\sqrt{\frac{\epsilon_0}{\mu_0}}\,\boldsymbol{a}_s \times \nabla_t E_{0s}\right\} \tag{H2}$$

The fields E_{0s} and H_{0s} can therefore be chosen real, and thus $\boldsymbol{E}_{0t}$ and $\boldsymbol{H}_{0t}$ will be imaginary. As a result Δ and P given in (9.39) and (9.40) are both real, as is then Δ/P.

H.2 Symmetry considerations

In the case when the waveguide is symmetric with respect to x, $[\epsilon(-x,y) = \epsilon(x,y)$ for all x and $y]$, we can always choose a given Cartesian field component (say E_{0s}) to be either an odd or even function of x. Maxwell's equations (9.30) then force all other field components to be odd or even functions as follows:

E_{0s}	even	odd
H_{0s}	odd	even
E_{0y}	even	odd
H_{0y}	odd	even
E_{0x}	odd	even
H_{0x}	even	odd

An inspection of the integrand in (9.39) shows that it is an odd function, and hence integrates to zero.

Appendix I

Asymptotic expansions

I.1 Asymptotic expansion of the ratio of two asymptotic series

In approximating (10.14) for large kR, we encounter the problem of finding the asymptotic expansion of the ratio

$$\frac{N(\nu_p, R)}{\partial/\partial\nu[D(\nu, R)]_{\nu=\nu_p}} \tag{I1}$$

where N and D have the known asymptotic expansions

$$N(\nu, R) \sim N_0(\nu) + \frac{1}{kR}N_1(\nu) + \ldots$$

$$D(\nu, R) \sim D_0(\nu) + \frac{1}{kR}D_1(\nu) + \ldots \tag{I2}$$

and ν_p is a root of

$$D(\nu_p, R) = 0 \tag{I3}$$

We must first obtain an asymptotic expansion for ν_p,

$$\nu_p \sim \nu_0 + \nu_1/kR + \nu_2/(kR)^2 + \ldots \tag{I4}$$

By formally substituting (I4) into (I3) and (I2), and taking a Taylor series expansion of the D_i about $\nu = \nu_0$, we find

$$D_0(\nu_0) = 0$$

$$\nu_1 = -D_1(\nu_0)/D_0'(\nu_0) \tag{I5}$$

and so on, by equating coefficients of like powers of $(kR)^{-1}$. A similar Taylor series expansion of $N(\nu)$ and $D'(\nu)$ then leads to

$$\frac{N(\nu,R)}{\partial/\partial\nu\,[D(\nu,R)]_{\nu=\nu_p}} \sim \frac{N_0(\nu_0)}{D_0'(\nu_0)} + \frac{1}{kR}\left\{\frac{\nu_1 N_0'(\nu_0) + N_1(\nu_0)}{D_0'(\nu_0)} - \frac{N_0(\nu_0)[\nu_1 D_0''(\nu_0) + D_1'(\nu_0)]}{[D_0'(\nu_0)]^2}\right\} + \dots \tag{I6}$$

Appendix J

Tuned window in rectangular guide

J.1 Mode expansions and matching

Fig. J.1 shows a symmetrical window of dimensions $D \times d$ in a guide of cross-section $a \times b$. An incident dominant mode is y-polarized and, particularly if d is small, there is little if any rotation of the plane of polarization of the resultant fields. This statement is exact both when $d = b$ and as $d \to 0$ and is probably a good approximation for all values of d. A discussion of the accuracy of this type of approximation in connection with a related problem is given by Mailloux,[1] and we shall adopt it here. It is equivalent to the assumption that the fields in the guide can be represented by longitudinal section (LS) waves[2] with zero E_x components.

For the electric fields on either side of the window we can take

$$E_y = \sin(\pi x/a)[e^{-jk'z} + Re^{jk'z}] + \sum_1^{\infty}\sum_0^{\infty}{}' A_{mn} \sin(m\pi x/a)\cos(n\pi y/b)\, e^{\Gamma_{mn}z}, \quad z<0$$

$$E_y = \sin(\pi x/a)\, Te^{-jk'z} + \sum_1^{\infty}\sum_0^{\infty}{}' B_{mn} \sin(m\pi x/a)\cos(n\pi y/b)\, e^{-\Gamma_{mn}z}, \quad z>0 \qquad \text{(J1)}$$

where

$$k' = (k_0^2 - \pi^2/a^2)^{1/2} = 2\pi/\lambda_g$$

$$\Gamma_{mn} = (m^2\pi^2/a^2 + n^2\pi^2/b^2 - k_0^2)^{1/2}$$

and Σ' means that the term $m = 1, n = 0$ is omitted from the double summation, as it has already been included in the first term.

At $z = 0$, $E_y = E(x, y)$ and has to be zero at the metal walls of the window. Working with this function as unknown, Fourier analysis gives

$$T = 1 + R = (2/ab) \int E(\xi, \eta) \sin(\pi\xi/a) d\xi d\eta$$

$$A_{mn} = B_{mn} = (2/ab)\epsilon_n \int E(\xi, \eta) \sin(m\pi\xi/a) \cos(n\pi\eta/b) d\xi d\eta \tag{J2}$$

where $\epsilon_n = 2$ unless $n = 0$, when it equals 1, and the integration is taken over the window opening.

The H_x components of the magnetic fields are

$$-k_0\zeta_0 H_x = k' \sin(\pi x/a)[e^{-jk'z} - Re^{jk'z}]$$

$$+j \sum_1^{\infty} \sum_0^{\infty}{}' A_{mn} \frac{(m^2\pi^2/a^2 - k_0^2)}{\Gamma_{mn}} \sin(m\pi x/a) \cos(n\pi y/b)\, e^{\Gamma_{mn} z}, \quad z < 0$$

$$-k_0\zeta_0 H_x = k'T \sin(\pi x/a)\, e^{-jk'z}$$

$$-j \sum_1^{\infty} \sum_0^{\infty}{}' B_{mn} \frac{(m^2\pi^2/a^2 - k_0^2)}{\Gamma_{mn}} \sin(m\pi x/a) \cos(n\pi y/b) e^{-\Gamma_{mn} z}, \quad z > 0$$

where $\zeta_0 = (\mu_0/\epsilon_0)^{1/2}$. (J3)

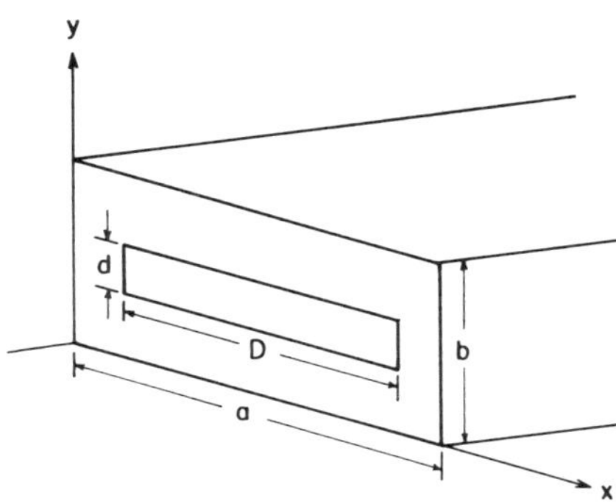

Fig. J.1 Window in rectangular waveguide

Equating the two values of H_x at $z = 0$ in the window aperture, dividing both sides by $(1 + R)$ and using (G2) to eliminate A_{mn} and B_{mn} gives

$$-jk' \sin(\pi x/a) R/(1 + R)$$

$$= \sum_1^{\infty} \sum_0^{\infty}{}' \epsilon_n (m^2\pi^2/a^2 - k_0^2) \sin(m\pi x/a) \cos(n\pi y/b) \times$$

$$\int E(\xi, \eta) \sin(m\pi\xi/a) \cos(n\pi\eta/b) d\xi d\eta \Big/ \left[\Gamma_{mn} \int E(\xi, \eta) \sin(\pi\xi/a) d\xi d\eta\right] \tag{J4}$$

Multiply this equation by $E(x,y)\sin(\pi x/a)$ and integrate with respect to z and y over the aperture. Remembering that a normalized susceptance jB across a line causes a reflection given by $2R/(1+R) = -jB$, we get

$$B = -\frac{2\sum_1^\infty\sum_0^{\infty}{}' \epsilon_n (m^2\pi^2/a^2 - k_0^2)\left[\int E(x,y)\sin(m\pi x/a)\cos(n\pi y/b)dxdy\right]^2/\Gamma_{mn}}{k'\left[\int E(x,y)\sin(\pi x/a)dxdy\right]^2} \tag{J5}$$

Now (J4) is an integral equation for $E(x,y)$ and (J5) is in variational form and is stationary for small variations of $E(x,y)$ about its correct value as given by (J4). We can accordingly approximate to B by assuming appropriate forms for $E(x,y)$, rather than attempting a solution of (J4) directly.

J.2 Variational approximation to the susceptance

As a first step, assume that $E(x,y)$ is the product of separate functions of x and y. This is certainly true when $d = b$ and also for d very small, and is consistent with the LS assumption made earlier. Then if $E(x,y)$ is replaced by $f(x)g(y)$ we get

$$B = B_0 + \sum_1^\infty B_n \tag{J6}$$

where

$$B_0 = -(2/k')\sum_2^\infty (m^2\pi^2/a^2 - k_0^2)^{1/2} \times \left[\int f(x)\sin(m\pi x/a)dx \Big/ \int f(x)\sin(\pi x/a)dx\right]^2 \tag{J7}$$

$$B_n = -(1/k')I_n \times \left[\int g(y)\cos(n\pi y/b)dy \Big/ \int g(y)dy\right]^2 \left[\int f(x)\sin(\pi x/a)dx\right]^{-2} \tag{J8}$$

$$I_n = 4\sum_1^\infty (m^2\pi^2/a^2 - k_0^2) \iint f(x)f(\xi)\sin(m\pi\xi/a)dxd\xi/\Gamma_{mn}$$

Before attempting to estimate a form for $f(x)$ we note that (J6) is exactly the equation that would have been arrived at if the problem of finding the susceptance of an inductive diaphragm had been treated by the variational method. Hence B_0 can be identified with the susceptance of an inductive diaphragm of opening D, and is given by[3]

$$B_0 = -(\lambda_g/a)\cot^2(\pi D/2a) \tag{J10}$$

(Actually, this is only the quasi-static approximation. Higher-order correction terms can be added if higher accuracy is required; though in view of subsequent further approximations this is probably not warranted.) At this point the quasi-static or other more exact forms for $f(x)$ and $g(y)$ could be inserted into (J7) and (J8) and the various quantities evaluated numerically. The disadvantage of this is the lack of analytical insight into the results. The trouble with using these exact forms analytically is the awkwardness of the resulting formulae. Accordingly we shall make rather crude estimates. For $g(y)$ we take the form that is analytically correct for narrow windows, and for $f(x)$ the form occuring at half-wave resonance. The latter is also correct at $D = a$, and since the range from $D = \lambda_0/2$ to a is usually not large the error in choosing a sinusoidal form is probably not great. For $g(y)$, however, the singularities at $y = \pm d/2$ are not needed at $d = b$, and hence some error for the very wide window is to be expected from these choices. The assumed forms accordingly are

$$\begin{aligned} f(x) &= \cos\left[\pi(x - \tfrac{1}{2}a)/D\right] \\ g(y) &= \left[1 - (b - 2y)^2/d^2\right]^{-1/2} \end{aligned} \tag{J11}$$

Much of the ensuing work comes from approximating integrals and summing the resulting series to get an analytically useful form.

J.3 Evaluation of I_n

We insert (J11) into (J9), use the addition formula for cosines to re-express the sine and cosine products, and change variables by putting $x - \xi = \sigma, x + \xi = a + \tau$. Then, on introducing the series

$$S(\sigma) = \sum_{m=1}^{\infty} \cos(m\pi\sigma/a)/\Gamma_{mn} \tag{J12}$$

we get

$$I_n = \iint \left[\cos(\pi\sigma/D) + \cos(\pi\tau/D)\right]\times$$

$$\left[(\partial^2/\partial\tau^2 + k_0^2)\, S(a+\tau) - (\partial^2/\partial\sigma^2 + k_0^2)\, S(\sigma)\right] dx d\xi \tag{J13}$$

Now $(\partial^2/\partial\tau^2)S(a+\tau) - (\partial^2/\partial\sigma^2)S(\sigma) = (\partial^2/\partial x\partial\xi)[S(x+\xi) + S(x-\xi)]$, and if $\cos(\pi\sigma/D) + \cos(\pi\tau/D)$ be temporarily re-written as $2\cos[\pi(x - \frac{1}{2}a)/D]\cos[\pi(\xi - \frac{1}{2}a)/D]$ then an integration by parts with respect to both x and ξ becomes possible. The integrated part vanishes at both limits on account of the form taken for $f(x)$, and the remaining terms give, on returning to σ and τ as variables, with $dx d\xi = \frac{1}{2} d\sigma d\tau$

$$I_n = (\pi^2/2D^2) \iint [\cos(\pi\sigma/D) - \cos(\pi\tau/D)] [S(\sigma) + S(a+\tau)] \, d\sigma d\tau$$
$$- \tfrac{1}{2}k_0^2 \iint [\cos(\pi\sigma/D) + \cos(\pi\tau/D)] [S(\sigma) - S(a+\tau)] \, d\sigma d\tau \qquad \text{(J14)}$$

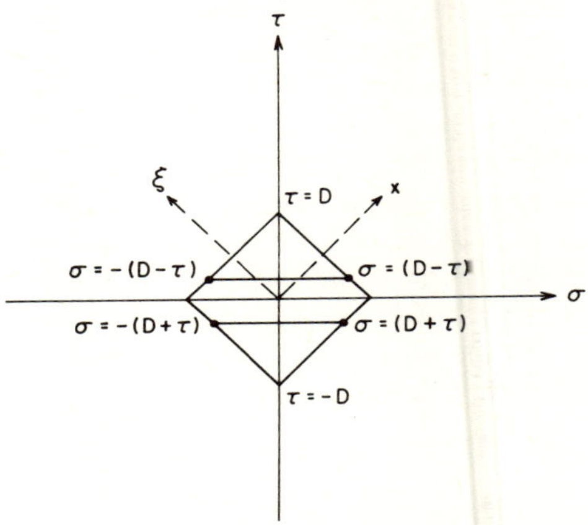

Fig. J.2 Change of variables and range of integration

The change of variables and range of integration is shown in Fig. J.2. The variable σ goes from $-(D-\tau)$ to $+(D-\tau)$, and τ from 0 to D in the upper half of the region. In the lower, σ goes from $-(D+\tau)$ to $+(D+\tau)$ with τ from $-D$ to 0.

Since, from its definition, $S(a+\tau) = S(a-\tau)$, the integrand in (J14) is symmetrical with respect to both σ and τ, and an integration in the first quadrant only can be taken, and the result quadrupled.

$$I_n = (2\pi^2/D^2) \int_0^D \int_0^{D-\tau} [\cos(\pi\sigma/D) - \cos(\pi\tau/D)] [S(\sigma) + S(a-\tau)] \, d\tau d\sigma$$
$$- 2k_0^2 \int_0^D \int_0^{D-\tau} [\cos(\pi\sigma/D) + \cos(\pi\tau/D)] [S(\sigma) - S(a-\tau)] \, d\tau d\sigma \qquad \text{(J15)}$$

Now it is readily verified, by integration by parts, that if $F(\sigma)$ is any suitable function of σ, that

$$\int_0^D \int_0^{D-\tau} F(\sigma) d\tau d\sigma = \int_0^D \tau F(D-\tau) d\tau \qquad \text{(J16)}$$

and

$$\int_0^D \int_0^{D-\tau} \cos(\pi\tau/D)F(\sigma)d\tau d\sigma = (D/\pi) \int_0^D \sin(\pi\tau/D)F(D-\tau)d\tau$$

Using these results to carry out the σ-integration in (J15) gives

$$\begin{aligned} I_n &= 2D(\pi^2/D^2 - k_0^2) \int_0^D \cos(\pi\tau/D)S(\tau)d\tau \\ &+ 2(\pi^2/D^2 - k_0^2) \int_0^D \tau\cos(\pi\tau/D)[S(a-D+\tau) - S(\tau)]\, d\tau \\ &+ (2D/\pi)(\pi^2/D^2 + k_0^2) \int_0^D \sin(\pi\tau/D)[S(a-D+\tau) - S(\tau)]\, d\tau \end{aligned} \tag{J17}$$

Now from Poisson's summation formula[5] we get

$$S(\tau) \sim -1/2\Gamma_n + (a/\pi)K_0(\Gamma_n\tau),\, \Gamma_n = (n^2\pi^2/b^2 - k_0^2)^{1/2} \tag{J18}$$

where the neglected terms are exponentially small, of order $K_0(\Gamma_n ma)$. These terms can generally be neglected for $m \geqslant 1$. Hence we can take

$$\begin{aligned} \int_0^D \cos(\pi\tau/D)K_0(\Gamma_n\tau)d\tau &\approx \int_0^\infty \cos(\pi\tau/D)K_0(\Gamma_n\tau)d\tau \\ &= \pi/2(\Gamma_n^2 + \pi^2/D^2)^{1/2} \end{aligned} \tag{J19}$$

[Actually, because of the neglected terms in (J18), the approximation in (J19) is rather better than it looks.]

J.4 Window susceptance

The integration in (J19) is really the dominant term of (J17), and before proceeding further we shall examine its contribution to $\sum_1^\infty B_n$. Since the denominator of B_n contains

$$\left[\int_{1/2(a-D)}^{1/2(a+D)} \cos(\pi x/a)\cos(\pi x/D)dx\right]^2$$

$= (4D^2/\pi^2)\cos^2(\pi D/2a)/(1 - D^2/a^2)^2$ the contribution to $\sum_1^\infty B_n$ is clearly

$$(k_0^2 - \pi^2/D^2)(\pi^2 a/4D)(1 - D^2/a^2)^2 \sec^2(\pi D/2a) \times$$
$$\sum_1^\infty \frac{[\int g(y)\cos(n\pi y/b)dy]^2}{[\int g(y)dy]^2 k'(\Gamma_n^2 + \pi^2/D^2)^{1/2}}$$

Now, apart from the initial factors, this is just the form that would have been arrived at if the problem of finding the susceptance of a capacitive diaphragm in a guide of breadth D had been treated by the variational method. It is therefore appropriate to identify this part with the susceptance of a capacitive diaphragm. Moreoever, at this stage, the analytically correct form of $g(y)$ can be used, giving the quasi-static result

$$\text{main part} \sum_1^\infty B_n = \frac{D}{a}\left[\frac{\pi(a^2-D^2)}{4aD\cos(\pi D/2a)}\right]^2 \times$$

$$\frac{1-\lambda_0^2/4D^2}{1-\lambda_0^2/4a^2}\cdot\frac{4b}{\lambda_g}\log\operatorname{cosec}(\pi d/2b) \qquad \text{(J20)}$$

(Note that λ_g here still refers to a guide of breadth a, since it arises from k'. Terms pertaining to a guide of breadth D only appear if higher-order corrections to the quasi-static formula are used.) The remaining terms in (J17) form a small correction to (J20). They are dealt with as follows. Use (J18) to replace the S-function in terms of $K_0(w) = \int_0^\infty e^{-w\cosh\theta}\,d\theta$. Interchange the order of integration so that the τ-integrations can be performed. A number of approximations may be made consistent with the quasi-static solution and the approximations used in obtaining (J18). Thus,

$$(D/\pi)\int_0^D \sin(\pi\tau/D)[K_0(\Gamma_n\tau) - K_0(\Gamma_n\overline{a-D+\tau})]\,d\tau$$

$$= (D/\pi)\int_0^D\int_0^\infty \sin(\pi\tau/D)\left[e^{-\Gamma_n\tau\cosh\theta} - e^{-\Gamma_n(a-D+\tau)\cosh\theta}\right]d\tau d\theta$$

$$= \int_0^\infty \left[1 + e^{-\Gamma_n D\cosh\theta} - e^{-\Gamma_n(a-D)\cosh\theta} - e^{-\Gamma_n a\cosh\theta}\right]\times$$

$$(\Gamma_n^2\cosh^2\theta + \pi^2/d^2)^{-1}\,d\theta$$

Neglecting the exponential terms in $\Gamma_n D$ and $\Gamma_n a$ and replacing $(\Gamma_n^2\cosh^2\theta + \pi^2/D^2)$ by $(n\pi/b)^2\cosh^2\theta$ gives the further approximation

$$(b^2/n^2\pi^2)\int_0^\infty [1 - e^{-\Gamma_n(a-D)\cosh\theta}]\operatorname{sech}^2\theta\,d\theta$$

$$= (b^2/n^2\pi^2)\left[1 - \int_{\Gamma_n(a-D)}^\infty dw\int_w^\infty du\int_0^\infty e^{-u\cosh\theta}\,d\theta\right]$$

$$= (b^2/n^2\pi^2)\{1 - K[\Gamma_n(a-D)]\} \qquad \text{(J21)}$$

where

$$K(v) = \int_v^\infty dw\int_w^\infty K_0(u)du$$

This function is shown in Fig. J.3. It approaches zero rapidly for quite moderate values of v. Since $\Gamma_n(a-d) \approx n\pi(a-D)/b$ it is seen that, even for quite small values of $a-D$, this term may often be neglected. For *very* small values, however, its value approaches unity, cancelling the first term in (J21). A quite good approximation, valid over the whole range, is

$$K(v) \approx e^{-v}(1+2v/\pi)^{-1/2} \tag{J22}$$

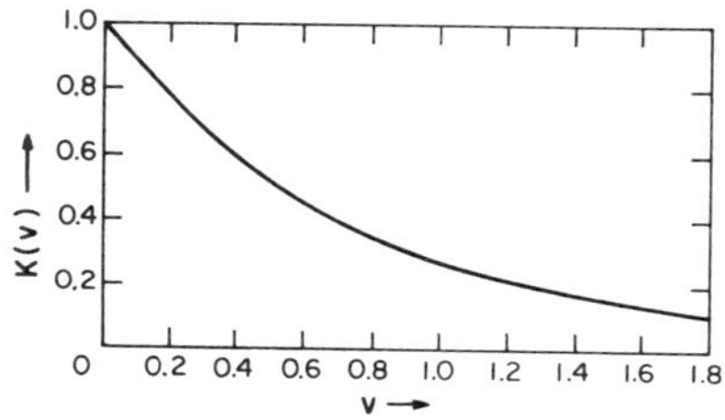

Fig. J.3 Function $K(v) = \int_v^\infty dw \int_w^\infty K_0(u)\,du$

The integral $\int_0^D \tau\cos(\pi\tau/D)[K_0(\Gamma_n\tau) - K_0(\Gamma_n\overline{a-D+\tau})]\,d\tau$ may be dealt with similarly and, with approximations of the same order made, the result turns out to be the same as (J21). Hence the correction to (J20) involves the summation

$$\sum_1^\infty \left[\int g(y)\cos(n\pi y/b)dy \Big/ \int g(y)dy\right]^2 \{1 - K[n\pi(a-D)/b]\}/n^2$$

Using (J11) for $g(y)$ these integrals can be expressed in terms of Bessel functions to give, since the odd terms integrate to zero, $\sum_1^\infty J_0^2(n\pi d/b)\{1 - K[2n\pi(a-D)/b]\}/4n^2$. Now the K term is often negligible, and the main effect comes from the first term. The series $\sum_1^\infty J_0^2(nw)/n^2$ can be summed. Starting with the Fourier series

$$\sum_1^\infty \frac{\cos nv}{n^2} = \frac{\pi^2}{6} + \frac{v^2 - 2v\pi}{4}, \quad 0 < v < 2\pi \tag{J23}$$

replace v by $u\sin\theta$ and integrate with respect to θ from 0 to $\pi/2$.

$$\sum_1^\infty \frac{J_0(nu)}{n^2} = \frac{\pi^2}{6} + \frac{u^2 - 8u}{8}, \quad 0 < u < 2\pi \tag{J24}$$

Now put $u = 2w\sin\theta$ and use the result $\int_0^{\pi/2} J_0(2w\sin\theta)d\theta = \frac{1}{2}\pi J_0^2(w)$. Then

$$\sum_1^\infty J_0^2(nw)/n^2 = \pi^2/6 + w^2/4 - 4w/\pi, \quad 0 < w < \pi \tag{J25}$$

Using this result and collecting all the terms together it is found that[6]

$$B = -(\lambda_g/a)\cot^2(\pi D/2a) + \left[\frac{\pi(a^2 - D^2)}{4aD\cos(\pi D/2a)}\right]^2 \times$$

$$\left\{\frac{1-\lambda_0^2/4D^2}{1-\lambda_0^2/4a^2}\cdot\frac{4bD}{\lambda_g a}\log\operatorname{cosec}(\pi d/2b)\right.$$

$$+ (\lambda_g/aD^2)\left[(b^2/3 + d^2/2 - 8bd/\pi^2) - 2(b^2/\pi^2)\times\right.$$

$$\left.\left.\sum_1^\infty J_0^2(n\pi d/b)K[2n\pi(a-D)/b]/n^2\right]\right\} \tag{J26}$$

The first term can be interpreted as the susceptance of the inductive part of the window, and the first part of the second term, involving the logarithm, as the susceptance of the capacitive part. It is seen to contain a factor $(1 - \lambda_0^2/4D^2)$, and this vanishes when the window is half a wavelength long. The total susceptance can only vanish if there is an infinite factor to counter it. Hence $d = 0$, for tuning with a half-wave window, a well-known result.

The remaining terms form a small correction to the rest of the equation. The quantity $(b^2/3 + d^2/2 - 8bd/\pi^2)$ does not quite vanish at $d = b$, as it should. This is because the form assumed for $g(y)$ is not quite appropriate to windows nearly as broad as the guide. For values of d greater than about $0{\cdot}8b$, at which the expression has a minimum barely above zero, this term may be ignored, along with the remaining infinite series whose first term also vanishes at this value. In any case, whatever the value of d, only the first few terms of this series are likely to be required.

References

1 Mailloux, R.J.: 'Radiation and near-field coupling between two collinear open-ended waveguides', *IEEE Trans. AP*, Vol. 17, No. 1, Jan. 1969, pp. 49–55

2 Lewin, L.: 'Theory of waveguides' (Newnes-Butterworths, London, 1975), p. 26

3 *ibid.*, p. 171, eq. 6.17

4 *ibid.*, p. 172, eq. 6.18, and p. 181, eq. 6.69

5 *ibid.*, p. 139, eq. 5.63

6 Lewin, L.: 'Advanced theory of waveguides' (Iliffe, London, 1951), p. 96, eq. 4.52. This original formula was in error by a missing factor (D/a) in the multiplier of the logarithmic term.

Index

£11.00